FOREWORD

AFTER NEARLY THREE DECADES of inattention, the North Atlantic is once again gaining recognition as a strategic space that is key to American and allied security. The United States Navy recently reinstated its Second Fleet, charged with conducting operations in the broader North Atlantic and all the way into the Barents Sea in the High North. At the same time, the North Atlantic Treaty Organization (NATO) is recovering some of its command structure to conduct operations in the North Atlantic, which will coincidentally be hosted in Norfolk, Virginia, home of the U.S. Navy's largest base and the site of NATO's now defunct Supreme Allied Command Atlantic (SACLANT). This is all in response to an aggressive Russia under Vladimir Putin, which has, among other things, annexed Crimea, threatened American allies, and sought to disrupt American democracy. In the maritime domain, Russia's new assertiveness is expressed through increased exercises, unsafe close encounters with U.S. and NATO member ships and aircraft, and the use of naval units to advance Russia's forays into the Middle East and elsewhere.

The U.S. Navy's leadership, from the Chief of Naval Operations to the commanders of U.S. Naval Forces Europe and the U.S. Sixth Fleet, have raised the alarm about the resurgent Russian navy and the threat it presents in the maritime domain in general and the North Atlantic in particular. And they are not alone. Gen. Philip Breedlove, USAF, who took over as NATO's supreme allied commander Europe and commander of U.S. European Command after my departure in

2013, has specifically noted the new challenges to allied security in the North Atlantic.

For me, this brings back memories of my time as a budding naval officer during the late Cold War, as I spent time on board surface combatants in the North Atlantic and the Mediterranean tracking and contending with Soviet submarines, surface ships, and aircraft. But the struggle over the North Atlantic predates the current contest with Russia and even the Cold War. Indeed, the North Atlantic is core to the Western world and central to the fortunes of both North America and Europe. America began as an Atlantic nation, and its independence was finally gained after the French navy broke the Royal Navy's control of the sea. The young American navy's first forays into the broader world was in the Atlantic as well, when President Thomas Jefferson sent its frigates across the North Atlantic to the Mediterranean Sea to fight in the Barbary Wars. Later, the North Atlantic formed a great highway for supplies and troops streaming from North America to Europe during both world wars. The North Atlantic convoys kept our European allies in the fight and helped determine the ultimate outcome of those two great conflicts. Indeed, Winston Churchill noted after the conclusion of World War II that "the Battle for the Atlantic was the dominating factor all through the war . . . everything happening elsewhere, on land, at sea, or in the air depended ultimately on its outcome." Geopolitically speaking, these two twentieth-century conflicts helped generate the concept of an Atlantic community of shared values and interests, which lies at the core of America's role in the world. Indeed, it is no accident that the world's great military alliance, NATO, has Atlantic prominently featured in its name.

This book serves as an important reminder of the importance of the North Atlantic to U.S. and European security and cohesion and how this maritime space has been used in conflicts that have determined major turns in recent history. But it also reminds us how the Atlantic has served as a space for close cooperation between the United States and its allies and partners, and even with its erstwhile adversaries at times. But this book is no mere history lesson.

The New Battle for the
ATLANTIC

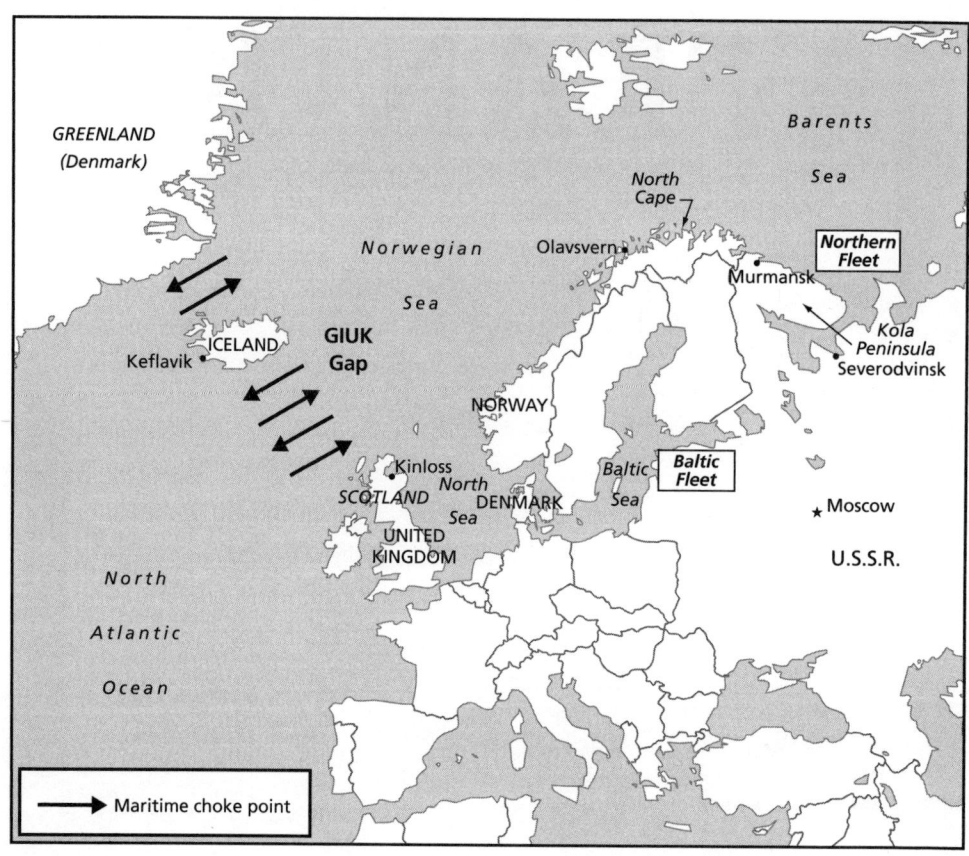

The New Battle for the
ATLANTIC

Emerging Naval Competition
with Russia in the Far North

MAGNUS NORDENMAN

Foreword by Adm. James G. Stavridis, USN (Ret.)

Naval Institute Press
Annapolis, Maryland

Naval Institute Press
291 Wood Road
Annapolis, MD 21402

Library of Congress Cataloging-in-Publication Data

Names: Nordenman, Magnus, author. | Stavridis, James G., writer of foreword.
Title: The new Battle for the Atlantic : emerging naval competition with
    Russia in the Far North / Magnus Nordenman.
Other titles: Emerging naval competition with Russia in the Far North
Description: Annapolis, Maryland : Naval Institute Press, [2019] | Includes
    bibliographical references and index.
Identifiers: LCCN 2018060114 (print) | LCCN 2019004006 (ebook) | ISBN
    9781682472842 (epub) | ISBN 9781682472842 (epdf) | ISBN 9781682472835
    (hardcover : alk. paper) | ISBN 9781682472842 (ebook)
Subjects: LCSH: Sea control—North Atlantic Ocean. | Sea-power—United
    States. | Sea-power—Russia (Federation) | North Atlantic Ocean—Strategic
    aspects. | North Atlantic Treaty Organization. | History, Naval—North
    Atlantic Ocean. | Cold War.
Classification: LCC V163 (ebook) | LCC V163 .N67 2019 (print) | DDC
    359/.03091631—dc23
LC record available at https://lccn.loc.gov/2018060114

27 26 25 24 23 22 21 20 19   9 8 7 6 5 4 3 2 1

First printing

Maps created by Chris Robinson.

To my father, Captain Jan Nordenman, Royal Swedish Navy (retired), who sparked my first interest in submarines and naval affairs and who did his part to keep the peace during the Cold War, in a sea not too far from the events described in this book.

CONTENTS

It paints a vivid portrait of how changing geopolitics, technology, transformed U.S. and allied naval forces, and a renewed Russian way of war will present a far different challenge for the United States and NATO leaders in the twenty-first century. It also offers a set of principles for the United States and NATO to consider as they prepare for the contest over the Atlantic. These principles rest on the lessons learned from the twentieth century but are also informed by these great changes we see playing out.

It is often said that this century will be a Pacific one. That is not untrue, but the North Atlantic is, once again, increasingly important as we seek to navigate these troubled waters in a new age of great-power competition.

Adm. James G. Stavridis, USN (Ret.)
Former Supreme Allied Commander at NATO, 2009 to 2013

PREFACE

THIS BOOK GREW not only out of a professional interest, but out of personal passion. I have never served in the navy, but sea service was never far away when I was growing up. My father was a career submarine officer in the Swedish navy and many school breaks were spent on base, where I would stay in my father's stateroom on the submarine squadron's command ship. I distinctly remember my fascination with the regular fire drills. I would sneak a peek through the hatch of my dad's stateroom at the crewmembers moving with a purpose through the ship while donning firefighting equipment. Later my father took the job of director of the new Swedish submarine program, and I spent time in the yard watching the massive sections of the submarine being fused together and equipment and cabling being installed by flinty and skilled yard workers.

As a teenager I joined the sea cadets, and I spent my summers rowing whale boats, practicing navigation, sailing, and signaling with semaphores and lamps. To say it made me a sailor would be a vast overstatement, but it did give me a small taste of sea life and an appreciation for the skill, knowledge, and hard work that goes into making a living at sea, whether commercially or in service of the nation. Permanent hearing loss from birth in one of my ears unfortunately put an end to my plans for a naval career, but it did not end my interest in naval and maritime matters.

As a proud American immigrant I am still, twenty years after coming to this country, astonished by the many opportunities this nation provides to those who come here. In my case it meant a chance to pursue, as a civilian, my interest in defense and national security, in particular my passion for the North Atlantic Treaty Organization (NATO) and maritime affairs. The Atlantic Council in Washington has been an excellent platform for me to do cutting-edge research on the most pressing national security issues of the day and to share the results with decision makers, senior military leaders, and civil servants with the U.S. government and its friends and allies. It has also allowed me opportunities to get under way for short periods on board U.S. Navy ships and with Marine units, as well as to visit ships and submarines of U.S. allies. One can learn a lot from books, memos, reports, and PowerPoint briefings, but there is no substitute for getting out to walk the ships, see the equipment in action, smell the smells, experience the environment, and, most importantly, talk to the sailors, Marines, and officers who make it all happen. I have always learned a great deal, and it has left me deeply impressed with those who serve at sea to keep America free and great, as well as those who serve alongside U.S. forces as allies.

Many thanks are in order to those who in various ways helped turn this book into a reality. I have benefited immensely from the advice and insights offered by a broad range of individuals during quick chats, long conversations, and interviews. Thanks are therefore in order to Sam LaGrone of the U.S. Naval Institute, my buddy from our days as cadets at the Virginia Military Institute, who introduced me to the editorial team at the institute and helped me pitch my book. Thanks to August Cole, author of the Navy cult novel *Ghost Fleet*, for his encouragement and advice on how to take on a writing project of this magnitude. Thanks to my many friends at the Norwegian Ministry of Defense who provided insights and regional context, among them Svein Efjestad, Arild Eikeland, Keith Eikenes, Per Kristian Krohn, and John Andreas Olsen. Thanks to Rear Adm. David Titley, USN (Ret.), former oceanographer of the U.S. Navy, for his thoughts and advice. Thanks to my boss at the Atlantic Council, Barry Pavel, who was supportive of this project from the get-go. Thanks to Bryan Clark at the Center for Strategic and Budgetary

Assessments in Washington for lending his expertise, and to Will Wiley, Atlantic Council navy senior fellow and U.S. submarine officer, for allowing me to bounce ideas around. Many thanks also go to Vago Muradian at Defense & Aerospace Report; Brad Peniston, the deputy editor of *Defense One*; Lee Willett from *Jane's Defence*; and many others. This book benefited in multiple ways from their collective insights, thoughts, and advice. Any mistakes or errors in this book are those of the author, of course. And last, but certainly not least, I need to thank my wife, Rita, who allowed me to sneak away to write while she corralled two energetic kids on her own. And to my daughter, Kate, and son, Bengt, who keep me grounded and remind me that the best job ever is being a dad.

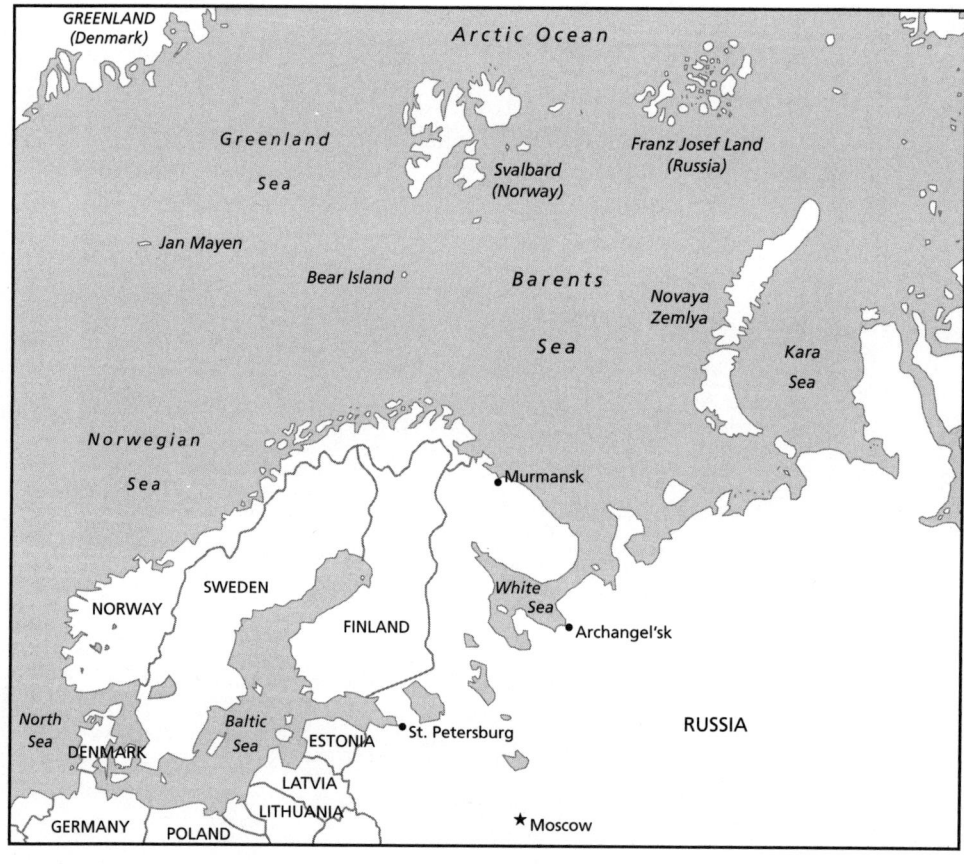

Introduction

On Tuesday, March 17, 2015, Scottish fisherman Angus MacLeod and his crew of four were out fishing haddock and skate off the Outer Hebrides in the *Aquarius*, a sixty-two-foot wooden trawler. The weather was cloudy and overcast, with drizzle throughout the day, classic Scottish weather in March. MacLeod was an experienced fisherman, having plied his trade off the coast of Scotland for more than thirty years. On this particular day he had put his nets out in waters that were more than 360 feet deep. Late in the evening his nets were suddenly and forcefully moved astern by a subsurface object. MacLeod increased the *Aquarius'* speed, and the crew worked feverishly to keep ahead of the net to avoid catching it in the propeller—something that could leave the *Aquarius* dead in the water. "The winch was under increasing strain as we tried to haul the rope. There was no way the net was snagged on the bottom—we were fishing well off the bottom. . . . Between the five of us, there is 110 years' experience and we have never experienced anything like that," the shaken skipper told the press. After roughly fifteen minutes of struggling to keep ahead, the dog-rope, which connects the net and boat, was cut by the *Aquarius'* propeller. The *Aquarius* was out of immediate danger, but the struggle with the net had left the rudder damaged, so MacLeod decided to launch the *Aquarius'* lifeboat to tow the *Aquarius* back to port. Something big and powerful had dragged the net along, and MacLeod swore it could not have been an animal. "It was not a whale—we have had whales in the nets before and the net is all twisted afterwards. Whatever it was, it was human powered—of that we are convinced," he explained.[1] That left a submarine the likely culprit.

Submarines do get caught in fishing nets from time to time. Modern fishing nets can be huge and can cover a significant space underwater to maximize the catch, while submarines are not the most nimble of craft, especially when operating close to the coast and in shallow waters. In 2016 a French trawler ensnared a Portuguese *Tridente*-class submarine at the western mouth of the English Channel, and in 1990 the British *Trafalgar*-class nuclear submarine HMS *Trenchant* dragged the trawler *Antares* below, with four of *Antares'* crew losing their lives in the accident. And the collision between the Japanese trawler *Ehime Maru* and the USS *Greenville*, which was demonstrating an emergency ballast blow for visiting dignitaries on board, killed nine crewmembers on board the *Maru* and led to a minicrisis in the U.S.-Japanese relationship.

But the incident with the *Aquarius* off Scotland was different. Safety at sea protocols call for the entangled submarine to quickly but carefully come to the surface to declare itself and to assist the fishing vessel that the net is attached to. This did not happen in the incident with the *Aquarius* off Scotland, and the British Ministry of Defense publicly stated that no Royal Navy or allied submarines were operating in the area at the time. Could it be that the Scottish fishermen had accidentally snagged a Russian submarine in their nets? This was, after all, around the same time that the Royal Navy had launched a hunt for a suspected Russian submarine close to the Faslane naval base in Scotland, and where London had requested assistance from maritime patrol aircraft (MPA) from Canada, the United States, and France to locate and chase away the intruding submarine. Later, the French navy reported a sighting of a Russian submarine in the Bay of Biscay, close to France's main basing for its nuclear submarines in Brest. Farther east, in the Baltic Sea, the Swedes launched a weeklong but ultimately fruitless sub hunt in October 2014 after credible sightings of a foreign submarine deep in the Stockholm archipelago. The Finnish navy dropped antisubmarine grenades while in pursuit of another suspected submarine outside Helsinki in the late spring of 2015.[2]

Together these incidents pointed to a worrying trend for Western admirals and decision makers in Washington, London, and elsewhere. The Russian submarine force, and the Russian navy more broadly, was back in the North Atlantic and operating more frequently and farther afield after almost three decades of decay and staying close to its bases on the Kola Peninsula in the Barents Sea. And this heightened Russian naval activity came with a backdrop of high tensions between America and its European allies on the one hand and Russia under president Vladimir Putin on the other. Putin and the Russians seemed bent on wrecking the European security order, a system that has been built and nurtured by the United States and European nations alike since the end of the Cold War, and on challenging the role of America as the ultimate guarantor of European peace and security.

Russia's behavior toward NATO, Europe, and the United States took an aggressive turn in 2008, when it fought a short but intense war with Georgia over the breakaway, Moscow-backed republics of South Ossetia and Abkhazia. Georgia is a small country on the eastern shore of the Black Sea and had traditionally belonged to Russia's so-called sphere of influence. Indeed, Joseph Stalin, the Soviet dictator who led the country to victory in World War II but who also killed millions of his own people, was born and raised in Georgia. Since the end of the Cold War, however, Georgia had doggedly pursued a course away from Russia and toward Europe and the United States and, it was hoped in Georgia's capital of Tblisi, eventual membership in NATO. Putin's war with Georgia in 2008 effectively froze Georgia's pursuit of NATO membership, which was likely Moscow's intent all along. But it did not make much of an impression in either Washington or in allied capitals in Europe. The intractable wars in Iraq and Afghanistan had become all-consuming affairs for the national security establishments on both sides of the Atlantic, and America was getting close to a contentious presidential election between Senator John McCain and a young, up-and-coming senator from Illinois by the name of Barack Obama.

But Washington and its European allies did notice when Putin launched his push to annex the Crimean peninsula from Ukraine, following the ouster of Ukraine's pro-Kremlin president Victor Yanukovych in early 2014. As of this writing, hostilities continue between Ukrainian forces and Russian-supported rebels (and sometimes Russian forces) in eastern Ukraine. But after 2014, Russian aggression was no longer confined to countries such as Georgia and Ukraine. Instead, since the beginning of the Ukraine crisis Russia has kept up a steady pace of large no-notice military exercises close to the borders of NATO allies Estonia, Latvia, Lithuania, and Poland, among others. Russia's air force has also been increasingly active around NATO's flanks, which has led to worrying close encounters with both commercial and military aviation, including a full Scandinavian Airlines flight leaving Copenhagen airport.[3] Russia has also built powerful networks of antiship, air defense, and ballistic missiles on the coasts of the Barents, Baltic, and Black Seas to make it difficult for the United States and NATO to quickly come to the aid of exposed allies in a crisis.

To date, NATO and the United States have responded to this new challenge to European security by deploying ground forces to Eastern Europe, increasing exercises, with a particular emphasis on ground operations, and upgrading infrastructure in the Baltic states and Poland to help speed reinforcements across the continent in case there is a crisis with Russia that demands a military response. But Russia's new ambitions are also expressed at sea, a domain that lends itself especially well to Russia's strategy of seeking to hold the United States and NATO at arm's length during a crisis, and using its military power to sense out weaknesses and test the resolve of decision makers. Driving a Russian tank across a NATO border would surely lead to escalation and quite possibly to war between Russia and the alliance. But the vastness of the sea, and the inherent challenge of controlling and monitoring it, gives Russia a range of possibilities for provocations, tests, intelligence gathering, deterrence, and preparing for operations, all the while being able to provide plausible deniability or ambiguity. And if a crisis comes, the Russian navy is increasingly well placed and equipped to operate in the far North Atlantic to strike at vital ports, airfields, and command-and-control centers that are needed to bring in U.S. and NATO reinforcements coming across the North Atlantic. If those cross-Atlantic reinforcements were stopped or delayed in coming ashore, NATO and the United States could very well lose a confrontation with Russia in Europe's east, far away from the shores of the Atlantic.

This means that the North Atlantic, and in particular the far North Atlantic, the Norwegian Sea, and the Barents Sea—all north of the maritime choke points between Greenland, Iceland, and the United Kingdom, in the maritime and national security community referred to as the GIUK gap—is also a crucial stage for the contest between Russia and NATO. Historically speaking, this is not new for the region. The North Atlantic was a central battleground during both world wars and the Cold War, as it is the connective tissue linking North America and the European continent. The D-Day invasion and the Battle of the Bulge may be more prominent in popular imagination about World War II, and the Berlin Wall and the hot wars in Vietnam and Korea may stand as key points in the Cold War, but the battles fought in and for control of the North Atlantic were crucial and at times no less deadly. Indeed, the battle for the North Atlantic during World War II stands as the longest continuous campaign of that war. The reason for its relative lack of prominence in the history books may just be a case of sea blindness. Humans, after all, live on land and have a natural propensity to orient themselves to their immediate environment. The North Atlantic region, on the other hand, is a maritime environment, vast, remote, and largely devoid of people, except on relatively small specks of land such as Iceland or along the Scottish and Norwegian coasts. One can easily visit the invasion beaches of Normandy, or even own a small piece of the Berlin Wall to keep on one's desk or bookshelf. The North Atlantic, however, is not as easy to take in, and it offers few of the telltale landmarks that suggest that events of importance took place there. But the fact remains that the North Atlantic is a crucial bridge that has shaped, and continues to shape, the destinies of both Europe and North America and has carried vast volumes of trade, people, and ideas between the two continents for centuries. But the North Atlantic has also carried armies and their supplies, as well as warships for conflicts both big and small. Indeed, it is no accident or mere happenstance that the most successful military alliance in world history, NATO, has North Atlantic in its name.

I decided to write this book to highlight the centrality of the North Atlantic in the competition between the United States and its NATO allies on the one hand and Russia under Vladimir Putin on the other. Since 2014 much has been written and said about the military risks to the Baltic states and the countries around the Black Sea from a newly assertive Russia, or the challenges of hybrid warfare for NATO, or the Russian use of information operations to stall decision making or, in the case of the 2016 U.S. presidential campaign, influence democratic elections. But the emerging contest between Russia and the West has important maritime components that to a large extent have been ignored as decision makers and military leaders focused on, among other things, bringing the U.S. Army back for exercises in Europe or countering Russian disinformation.

This new contest between NATO and Russia, and the re-emerging potential for war in Europe involving the great powers, returns the North Atlantic to its strategic role as the maritime superhighway between Europe and North America, a key factor in the ultimate victory in both world wars and the Cold War. The previous contests over control of the North Atlantic shared some enduring factors, but they were also different, depending on the actors in the contest, the expanse of the contest, the technologies deployed, and the strategies employed. And so it is with the emerging battle for the Atlantic that this book focuses on, which contains echoes of the struggles that came before, but also has its own twenty-first-century quality.

This book is intended to reintroduce the North Atlantic and its role in security to a broader audience and is not meant as a narrow or in-depth treatment about naval combat and its associated ships and technologies. Instead, this book combines themes on naval affairs, technology development, geopolitics, and the future of NATO in the twenty-first century. All these factors, and how they have historically functioned and interacted in the Atlantic, need to be considered as the United States and its NATO allies once again prepare to defend the North Atlantic in the twenty-first century.

CHAPTER 1

An Introduction to the North Atlantic

THE NORTH ATLANTIC is a distinct maritime region and basin that forms the upper half of the Atlantic Ocean, the second largest ocean on the planet after the Pacific. The North Atlantic covers some sixteen million square miles of open ocean, nearly 12 percent of the world's total ocean surface. The North Atlantic also provides connections to a wide range of seas and oceans, including, in the north, the Norwegian Sea, the Barents Sea, the Baltic Sea, and the North Sea, along with the Arctic Ocean, and in the south to the Mediterranean, the Black Sea, and, through the Suez Canal, the Red Sea. The western and middle part of the North Atlantic is broadly open until one arrives at the many deep bays of the east coast of North America, while the eastern North Atlantic is more constrained by the choke points formed by Greenland, Iceland, and the United Kingdom (Scotland), the so-called GIUK gap. A collection of smaller islands, such as the Faroe Islands, the Shetland Islands, and Jan Mayen provide additional dry land toeholds north of the GIUK gap. A second choke point can be found farther north, between the Svalbard Islands and the Norwegian mainland, which provides the entrance to, or exit from, the Barents Sea.

This maritime geography of the North Atlantic is largely lost on today's general public in North America and Europe. People rarely cross the North Atlantic via ship anymore. Indeed, most people come into momentary contact with these geographical features from a screen mounted on the back of an airplane seat, as the map there shows their airliner treading its way across the North Atlantic from North America to Europe or the other way around. But even then the geography is barely understood, as it fails to give proportion to the North Atlantic region; a jet will travel across the North Atlantic gap in some five to six hours, but the space below is vast. But while the flow of people by sea across the North Atlantic has dwindled in the jet age, other flows have actually intensified, including trade, energy, and information, with the latter flowing across the North Atlantic Ocean floor in submarine cables. Many of America's largest and busiest ports can be found on the Atlantic coast, including Port Newark, Hampton Roads, Baltimore, and Charleston. On the other side of the North Atlantic, European megaports servicing the Atlantic and the broader global maritime domain include Rotterdam, Antwerp, Hamburg, and Bremerhaven. All in all, North America and Europe trade across the North Atlantic for well above a trillion dollars each year. Indeed, even with the growing economies of the Asia-Pacific region, the U.S.-European trading relationship remains the largest in the world. And new trade linkages across the North Atlantic are currently being opened up due to America's emergence as an energy exporter. In June 2017 the first liquified natural gas (LNG) tanker departed a port in Louisiana for its trek across the North Atlantic to Poland.

But while trade relationships are key, they should not obscure the North Atlantic's role as the strategic connector of two great continents, which indeed forms the great core of the Western world. This is often lost on a modern world that is more inclined to consider sea spaces as barriers rather than bridges. This forgetfulness about the role of the sea is unfortunate in an age of rising geopolitical competition and stirrings that the age of Western global leadership, centered around the North Atlantic, and the liberal world order that it has built, may be coming to an end.

The North Atlantic connection has inevitably shaped both Europe and North America economically, politically, socially, culturally, even gastronomically. The North Atlantic maritime domain has also directly influenced wars and conflicts that have shaped the world we live in. Indeed, much of the competition among the European colonial powers played out in the North Atlantic, as they sought to protect their trade, and disrupt that of the others, using sea power. The United States' own national development was in no small part dependent on the North Atlantic. The independence of the United States was finally made a reality with the help of French sea power, which broke the British control of the North American east coast. The American Civil War was also fought in the North Atlantic, as the Confederacy sought supplies and support from European nations, and the Union worked to blockade the South's ports from receiving badly needed arms and from shipping out the Confederacy's cash crop, cotton, to profitable European markets.[1]

But at no time was the North Atlantic so intensively contested as during the relatively recent past, with the world wars in Europe and the Cold War. The intensity of warfare in the North Atlantic was driven by the fact that the world wars were truly monumental conflicts that harnessed nearly all the resources of the combatant nations. The wars also drove the introduction of new technologies and their further development, such as airplanes and submarines, that remain with the world in more advanced forms today. The Cold War, meanwhile, saw little in the way of actual fighting in Europe and the North Atlantic, while hot wars raged in Asia and the Middle East. But in the North Atlantic, vast resources were poured in and advanced technologies developed by the two superpowers and their allies to prepare for a potential clash of arms at sea as part of a larger conflict between the two superpowers.

The major power conflicts in Europe during the twentieth century were not fought over the North Atlantic per se. They were clashes over the control and future destiny of the European continent and its role on the global stage. The North Atlantic domain, however, loomed large in each conflict as the bridge for supplies and troops that enabled the combatants to fight on in a struggle that would have quickly drained the resources of an isolated European continent or Atlantic nation. This brings home the fundamental point about the enduring role of the North Atlantic for both American and European defense and security: a war in Europe will not be won in the North Atlantic, but one can surely be lost there.

CHAPTER 2

Submarines

A Primer

THIS IS NOT A BOOK ABOUT SUBMARINES. They do, however, play a major role throughout this book as an evolving type of naval vessel that has played a key role in the struggles for control of the North Atlantic. A short primer on the submarine as a system and its place in naval strategy and operations is therefore in order.

A submarine is a vehicle able to conduct independent operations under the surface of the sea. While rudimentary submarines were introduced during the nineteenth century and even played a bit role in the Confederate navy during the American Civil War, submarines really came to the fore as an effective weapon of war during the twentieth century. Submarine propulsion is provided by diesel engines when the submarine operates on the surface, while batteries provide power, but normally lower speeds, for subsurface operations. Submarine batteries provide quiet propulsion but also generate the need for relatively frequent surfacing, or snorkeling, to recharge the batteries using the diesel engines. This process creates a real vulnerability for conventional submarines and negates the submarine's real advantage: stealth. In the late 1950s, submarines with nuclear propulsion began to emerge, first in the United States and later in the Soviet Union, the United Kingdom, and France. Nuclear propulsion allowed for longer periods of submersion and greatly extended the operational reach of the submarine forces. Effective quieting of nuclear submarines, however, remained a challenge for many years.

Modern submarines use sonars as their primary sensors to gain situational awareness of their surroundings and to detect, track, and attack targets below and on the surface. Torpedoes have been the mainstay weapon of submarine forces, but many current classes of submarines can also lay mines and fire antiship and land-attack cruise missiles. The nuclear powers, such as the United States, France, the United Kingdom, China, and Russia, also rely on submarines as a highly survivable platform for their nuclear deterrent capabilities. Strategic nuclear ballistic missile submarines (SSBNs) are built to serve as a platform for nuclear strikes from the sea and are ill-equipped for a subsurface attack role, although they can normally defend themselves with torpedoes if the need arises. In the case of the United States, France, Russia, and China, the SSBNs form a leg in the nation's nuclear deterrent, along with aircraft and missile-borne nuclear weapons. In the case of the United Kingdom, the SSBNs are the main part of the deterrent force. Submarines capable of firing cruise missiles can also take on a nuclear mission, as many types of cruise missiles can accept both conventional and nuclear warheads.

Submarines have rarely fired their torpedoes in anger against surface ships or submarines since World War II. But in the post–Cold War era submarines have arguably been busier than ever in other roles. The United States and the United Kingdom have used submarines on a number of occasions to strike targets ashore in places such as Iraq, Afghanistan, and Serbia. Modern submarines are also uniquely capable intelligence gatherers, as they can approach foreign coasts or operate in sensitive areas without being detected and can, among other things, monitor ship traffic or naval operations and exercises, collect electronic intelligence, or intercept communications from the many submarine cables that crisscross the world's oceans. Indeed, the performance of this mission constituted some of the most consequential intelligence successes for the United States during the Cold War.[1] They can also be used to covertly insert and extract special operations forces on hostile beaches.

Submarines, however, are not without their own weaknesses and drawbacks. For one, conventional submarines are slow while submerged and can easily be outrun by their targets on the surface, or hunted down by faster moving antisubmarine warfare (ASW) frigates, helicopters, and maritime patrol aircraft (MPA). Nuclear submarines, on the other hand, can certainly make better speed underwater, but higher speed leads to more noise and possible detection by ASW sensors. When traveling at high speeds, the submarine's own sensors also tend to become less effective. This means that a submarine normally must position itself ahead of a surface force or submarines that it seeks to attack or surveil. This can be a painstaking, slow, and deliberate process, and one that must nearly always be aborted if the submarine is detected during the effort. The range of the main sensors on board a submarine, passive and active sonar, is also fairly limited when compared to those carried by surface ships. Even modern submarines with towed sonar arrays cannot consistently track targets that are much farther away than twenty nautical miles. Additional sensors, such as radar and electromagnetic detectors, can be deployed on top of masts, but that means breaking the surface, and modern naval radars on board surface warships and aircraft can easily detect the masts that the submarine sends up. Torpedoes, the traditional main armament of the submarine, are powerful weapons that can break the back of even the largest surface ships, unlike most cruise missiles. The range of torpedoes, however, is relatively short, with the effective range of the current U.S. standard torpedo, the Mark-48, around twenty nautical miles. This means that a submarine using torpedoes must get relatively close to its intended target. Furthermore, if detected and attacked, submarines have little in the way of self-defense weapons. Modern submarines can launch decoys, and experimental designs for submarine-launched air-defense missiles against ASW helicopters have been developed. Still, this is far from adequate for the submarine to fight back against ASW forces. In short, a submarine must break off the attack or approach to make its escape once it is detected and tracked, as there is little it can do against antisubmarine weapons.[2]

The submarine rose to prominence during the conflicts of the twentieth century, and, along with naval aviation, helped change warfare at sea forever, not least by making very large surface combatants, such as the battleship, largely obsolete.[3] The transformative power of the submarine comes in no small part from the fact that the submarine enables strategically weaker, but not necessarily unsophisticated, opponents to challenge far larger and more powerful navies in ways previously unthinkable. In important ways it has become the defining naval system of the militarily inferior. But while the weak have sought to build submarine forces to offset the advantages of stronger opponents, it does not mean that the strong have not found uses for submarines too.

But to say that the submarine is the weapon of the weak is not to say that submarines are crude instruments of war. Quite the contrary. Modern submarines are marvels of engineering and systems integration, with striking similarities to space vehicles in terms of complexity and exacting demands. Indeed, both submarines and space vehicles operate in forbidding environments where a system failure can rapidly turn into a complete disaster with the destruction of the submarine (or space vehicle) and loss of all hands. Few nations today can be said to be able to independently design and construct submarines, whether conventional or nuclear powered. The independent submarine construction club currently includes only the United States, the United Kingdom, France, Germany, Japan, Sweden, Russia, and China. And once they are built they must be maintained and periodically modernized, an accomplishment that itself has proven difficult for even advanced nations. In short, if a nation can design, build, and maintain submarines, it belongs to a small and exclusive group of the world's most technologically sophisticated and competent countries. To successfully operate and maintain a submarine force means that a nation has truly arrived on the scene as a credible military power. It is therefore not surprising that the eager, ambitious, and emerging nations of Asia, from China and India to Singapore and Indonesia, are spending a major share of their growing defense investments on undersea warfare.

Subsurface warfare, ASW, and high-end naval warfare in general between two capable opponents has been a rare thing since World War II, but one of the few modern examples of high-intensity naval combat, the Falklands War between Argentina and the United Kingdom in 1982, points to the effectiveness of submarines. Even a small number of submarines held by the adversary can be intimidating for national leaders as well as seagoing military commanders.

During the short and sharp Falklands War between Argentina and Britain in 1982, the Argentinian Navy only had three German-built 209 boats (a relatively new design at the time) in the fleet, with only one, the *San Luis*, out at sea after the sister boat *Santa Fe* was destroyed by the Royal Navy early in the conflict. Nevertheless, the threat of this one diesel-electric submarine helped keep the Royal Navy's two aircraft carriers at a significant distance from the Falklands, which impacted the ability to reach the islands with naval aviation. The presence of the *San Luis* also tied up a number of Royal Navy warships that were dispatched to hunt for her. The Royal Navy sub hunters also expended a large number of torpedoes, some reports say at least fifty, and depth charges in their attempts to sink the *San Luis*. The Royal Navy ASW efforts ultimately failed to destroy the *San Luis*, but they did eventually force her away from the British task force and back into an Argentinian port. But the example of the *San Luis* also points out that effective subsurface warfare requires more than just a sophisticated submarine. While the *San Luis* proved to be a headache for the Royal Navy, the submarine failed to score any torpedo hits against any Royal Navy ship, in spite of several attacks. This was in no small part due to the inexperience of the Argentinian crew and a lack of maintenance of the torpedo system on board the submarine. Thus, without skilled and well-trained crews who are intimately familiar with their submarines and up-to-date maintenance, the submarine itself is only so much steel in the water. Of course, submarines proved intimidating to the Argentinian Navy as well. After the Royal Navy *Churchill*-class nuclear attack submarine the HMS *Conqueror* sank the destroyer *General Belgrano*, the Argentinian surface fleet vacated the sea and remained in port until the end of the war. The Royal Navy submarines involved in the Falklands War also provided incredible amounts of intelligence gathering and reconnaissance that helped guide the British landing force.[4]

In an age of operations in cyberspace and the increasing military use of robotics, the submarine remains alive and well in navies around the world, from small coastal forces to major powers with reach across the oceans. The submarine, both conventional and nuclear powered, continues to evolve and is set to add new capabilities in the near future, including the use of unmanned systems. Little suggests that the submarine will play less of a role in future crises and conflict in the maritime domain in the twenty-first century. This is no less true in the emerging competition between the United States and its European allies and a revanchist Russia in the North Atlantic.

PART I

War

AS THE MARITIME BRIDGE between North America and Europe, the North Atlantic has played important roles in major conflicts throughout the twentieth century. Plenty of books have been written about the struggles at sea during the two world wars and the Cold War, so there is little need here to do another deep dive on the conflicts at sea. Still, it is important to note the highlights of each conflict, as they reveal the enduring importance and dynamics of the broader North Atlantic region, as well as the set of factors that were and remain crucial to the currently emerging maritime competition in the region. Twentieth-century warfare in the North Atlantic also taught the combatants valuable lessons about the technologies, tactics, and alliances needed to succeed in the Atlantic. The history of twentieth-century warfare also points to the price to be paid when a combatant is poorly prepared or gets the nature of the challenge wrong. This section of the book concludes with a summary of lessons learned and key factors drawn from each battle for control of the Atlantic that remain relevant today and into the future.

CHAPTER 3

World War I

The First Undersea Contest in the Atlantic

Strategic Setting

IN THE LATE SUMMER OF 1914, many assumed that World War I, or the Great War, would turn into a short and sharp conflict. Enthusiasm was high in most capitals. The German grand strategy, articulated in the Schlieffen Plan, called for a rapid advance through Belgium and into France, where the German forces would overcome the French and take Paris before British reinforcements could arrive. But the western front bogged down into a war of attrition and trenches with the German defeat at Marne in September 1914. Meanwhile, the eastern front remained fluid and the ground war raged over vast distances, but it still was clear that there would be no quick victory in the east either. As it became understood that the war would be a long-term effort, Germany began to consider how it could knock Britain out of the war by closing off its access to outside supplies, ranging from food to fuel, something the island nation was absolutely dependent upon. Britain's Royal Navy was superior in surface warships and had put an end to large surface actions by the German navy with the Battle of Jutland in June 1916. But the recently introduced submarine would give Germany another opportunity to threaten the sea-lanes and the shipping going to and from British ports.

The Coming of the Submarine

As World War I opened in Europe, the submarine was a relatively new type of platform that only recently had been introduced into the navies of the major powers. The U.S. Navy had formally commissioned its first submarine in 1900, while the Royal Navy and the German navy introduced their first submarines in 1901 and 1906, respectively. The submarines in service just before and into World War I had short ranges (since they used gas rather than diesel engines) and had limited endurance below the surface. They were, for all intents and purposes, submersibles rather than real submarines. But with the Royal Navy surface fleet showing its dominance and the ability to keep the German surface fleet close to home ports, the idea emerged to use German submarines to blockade the British and put immense pressure on the British economy. At the time, this was considered an innovation in warfare at sea, as submarines had originally been conceived by naval planners and strategists as a complement to and providing support for the great surface battleships. But during exercises and tests it became evident that the submarines were too slow to keep up with the powerful battleships and destroyers, and the limited means for fast communications made effective operational coordination nearly impossible. It was also soon discovered that submarines did not fare particularly well in combat against surface warships either, in no small part due to the superior speed of the surface ships. The submarines were thus found to be better suited as independent commerce raiders.[1]

The use of submarines in naval warfare, especially against merchant ships, was not uncontroversial. Naval and national leaders at the time considered the submarine an ungentlemanly weapon that gave unarmed merchantmen little warning or much to fight back with. Indeed, the discussions around the ethics of submarine warfare before and during World War I is not unlike many of the discussions that arose regarding the use of unmanned systems in a combat role, such as the Predator and Reaper unmanned aerial vehicles, which leave the operator safe at home and far from the field but give the target little notice and virtually no chance of fighting back.[2]

The Birth of Antisubmarine Warfare

The British response to the German submarine threat around the British Isles and in the North Atlantic was halting and characterized by trial and error. ASW weapons were rudimentary and were sometimes little more than small explosive charges, mounted on top of a handle, that were tossed by hand into the water above a suspected submarine. Another favored ASW tactic was to simply ram a detected submarine on the surface, which was far less maneuverable and much slower than the oncoming surface warship. Even larger unarmed merchant and passenger vessels were advised by the Royal Navy to, if attacked, go to flank speed and attempt to run over the submarine if it was on the surface.[3] Detection and targeting methods were rudimentary too, consisting of little more than the visual identification of surfaced submarines using binoculars, followed by the announcement via radio of the sighting to nearby surface warships patrolling the area.

But while simple in nature, British ASW efforts during the Great War turned out to be costly affairs in terms of time and valuable resources spent, a dynamic that has proven timeless in the ASW field. In one particular ASW operation in September 1917, no more than three German submarines were able to sink thirty cargo ships in an area that was under surveillance by approximately two hundred ships of various classes, ranging from destroyers and torpedo boats to armed auxiliary ships that had been pressed into the war effort, while the submarines themselves were directly pursued by twenty destroyers and armed merchantman.[4] All in all, by 1917 around two-thirds of the Royal Navy's destroyer force, and all its submarines and mine sweepers, were involved in the defense against Germany's submarine force, drawing them away from other missions that may have advanced Britain's war aims at sea.[5]

It is easily overlooked that the point of ASW is not necessarily to sink the submarine, but only to stop it from carrying out its purpose, although the destruction of the submarine would certainly fulfill that. Sometimes strategies that either frustrate the submarine's operations or stop the submarine from reaching its intended area of operations can be just as effective. This approach was also tried during the Great War with the so-called Dover Barrage, a chain of mines and steel netting strung across the English Channel from Dover on the coast of southern England to Belgium. Initially this approach was successful and kept German submarines out of the English Channel. Later in the war, however, the German fleet discovered that the British could not provide effective around-the-clock monitoring of the chain of mines and netting and that submarines could cross the barrier at night while transiting on the surface.

Letting the Leash off the Submarines

While the German U-boats proved to be hard to detect and sink for the Royal Navy during the early part of World War I, the dent that the German submarines were able to make in the rate of shipping into Britain was not very impressive either. The sinking of merchant vessels was simply not at a rate that British and neutral shipping could not make up. This was largely due to the relatively small number of submarines the German navy had at its disposal during the early part of the war, along with the stock of prewar torpedoes that often failed to leave the torpedo tube when fired or detonate when contacting the target ship.[6] The technological effectiveness of the German submarine campaign steadily improved, with more available submarines and a new type of torpedo, but the German efforts in the North Atlantic and the North Sea still fell far short of the 600,000 tons of shipping sunk per month that was thought to be needed to truly impact Britain's ability to keep fighting the war.

The failure of the German submarine offensive of that period in the war was due to the prize regulations, which directed naval vessels stopping civilian shipping vessels on the high seas to search them and sink them only after the crew and passengers had been evacuated. These regulations were extended to submarines at the 1909 London Conference. But U-boats were rarely in a position to pursue and overtake the faster merchantmen, and the time it took to search and evacuate the merchantmen left the submarine dangerously exposed on the surface to detection and attacks by ASW ships. This practice became especially dangerous for German submarines after the British introduced the Q-ship, heavily armed merchantmen that were more than capable of seriously damaging or even sinking a submarine in a fight. The size of the German submarine force could be helped with increased submarine yard production and the torpedo type could be replaced with newer models, but the prize regulation challenge could only be overcome through a change in strategy. Indeed, the British use of armed merchantmen formed part of the German rationale for abandoning their adherence to the prize regulations. Germany's unrestricted submarine warfare led to one of the most dramatic events at sea during World War I: the sinking of the *Lusitania*.

The *Lusitania* left New York on May 1, 1915, and headed into the North Atlantic toward her destination of Liverpool. Launched in 1906 and for a time the world's largest ocean liner, by 1915 the *Lusitania* had already made some two hundred crossings of the Atlantic. The threat to Atlantic shipping was publicly known as the *Lusitania* headed away from New York, and the German embassy in Washington had even taken out advertisements in major American newspapers giving notice that even American ships heading for Britain would be at risk of sinking by German submarines as they approached Europe. Indeed, when the Great War broke out the *Lusitania* had been painted in dull and dark colors to make spotting her out in the North Atlantic more difficult. By May 1915, however, *Lusitania*'s color scheme had been reversed back to its original lighter palette as the ship headed into the Atlantic. Many civilian ships had been drawn into war service by the British government, and the *Lusitania* was formally on the rolls of the British fleet reserve. But unlike many of the ships of her kind she remained in commercial passenger service across the North Atlantic. And this particular trip was to be a tight squeeze for the passengers on board the *Lusitania*; just short of two thousand passengers and crew were on the ship on this journey to Liverpool.

A few days before *Lusitania*'s departure from New York, the German submarine *U-20* left its base in Borkum and headed into the North Sea. Handed over to the German Imperial Navy in 1912, *U-20* was part of the small submarine force that Germany had at its disposal at the beginning of World War I. By April 1915 *U-20* had already performed a number of patrols and had sunk six British merchantmen since the beginning of the war. *U-20* had also been in tough spots during its combat patrols. In the English Channel *U-20* had become entangled in one of the antisubmarine nets emplaced by the British and was forced to the bottom, but she managed to extricate herself and escape the hovering Royal Navy frigate that had observed *U-20* getting tangled in the net. Armed with six torpedoes and with a crew of thirty-five, *U-20* headed toward the Irish Sea, directed there by the navy command ashore, which was able to decipher British communications and provide a general idea of British merchant ship traffic patterns.

After taking up its station off the southern coast of Ireland, *U-20* managed to find and sink three British merchantmen on May 5 and 6 but also saw another attack foiled by a failing torpedo. Later, *U-20* also encountered a probable British Q-ship but broke off the attack before the Q-ship discovered the submarine. *U-20* spotted the *Lusitania* in the early afternoon of May 7, as the submarine's patrol was drawing to a close, and maneuvered into a position within eight hundred yards of the liner. The submarine's commanding officer noted that the *Lusitania* was clearly a passenger liner rather than a merchant ship, but he also knew that the *Lusitania* belonged to the British fleet of reserve ships. *U-20* fired a single torpedo from periscope depth, which traveled the short distance to the *Lusitania* at a depth of ten feet and impacted the hull under the bridge. The demise of the *Lusitania* was quick. After the torpedo detonated, a secondary explosion shortly followed, likely originating with the ship's steam boiler or from coal dust. Of the close to two thousand passengers and crew, only some seven hundred survived the attack and the long stay in the lifeboats before aid from the Irish coast could reach them. *U-20*, meanwhile, observed the sinking of the liner for a short while and then dove to eighty feet and left the area. *U-20* would go on to perform another four patrols against British and allied shipping, sinking some thirty additional ships before it was beached and damaged beyond recovery in late 1916. *U-20*'s commanding officer, Lieutenant Walther Schweiger, would later perish, along with the rest of his crew, on board *U-88* in late May 1917, which was sunk after striking a British mine while being pursued by a Royal Navy sub hunter.

The sinking of the *Lusitania* killed 128 Americans and caused an uproar across the United States. The sinking was followed by additional attacks of liners that killed more Americans, which further increased the ire among the American public. The American reaction to the *Lusitania* contributed to the German decision to cease attacking ocean liners plying the North Atlantic until the reintroduction of unrestricted submarine warfare by Germany in 1917.

After a pause, due to American pressure after the sinking of the *Lusitania*, Germany intensified its unrestricted submarine warfare campaign in January 1917, and it expanded its U-boat campaign to include the Barents Sea as a new northeastern boundary. As 1917 progressed, the campaign would be further extended to encompass most of the North Atlantic, far beyond the campaign's original extent, which had been limited to the North Sea and the English Channel.[7] At the same time, additional German U-boats sat ready at their piers to join the expanded effort. The renewed German focus on submarine attacks against both British and neutral shipping had a substantial strategic effect. Shipping losses increased noticeably, and in April 1917 more than 800,000 tons of shipping was sunk by German submarines, a rate well above what was believed the United Kingdom could replace. All in all, during 1917 German submarines sank more than five million tons of shipping, a sink rate that far overtook the rate of new construction of merchant vessels.[8] Some estimates indicated that by mid-1917 Britain had mere weeks before it ran out of key food staples; and serious discussions were held in London about whether the effort to support the ground war on the European continent would have to be scaled back to divert shipping to bring grain from North America.[9]

By the middle of 1917 Britain seemed to be on the path to starvation under the pressure applied by the expanded German submarine campaign, which continued in spite of the vast resources the Royal Navy had thrown at guarding the sea-lanes. The creation of a convoy system took some of the pressure off, even though the introduction of the convoying method was at first strongly opposed by the leadership of the Royal Navy.[10]

While the introduction of the convoy system proved an effective operational countermeasure to the broadening German U-boat campaign in 1917 and 1918, the new German aggressiveness below the surface of the North Atlantic also proved to be strategically self-defeating: The public reaction to the sinking of American ships and the loss of American life contributed to bringing the United States into the war. This tilted the balance on the western front, making victory there an impossibility for Germany and its allies.

Russia and the First Battle of the Atlantic

The battle for control of the maritime domain during World War I occurred primarily in the North Sea, the English Channel, and the North Atlantic south of the GIUK gap. Indeed, the *Lusitania*, which had become such a symbol of the cost of the war at sea during World War I, happened in the Irish Sea. All was not quiet, however, north of the GIUK gap and in the waters approaching northern Russia. The U.S. Navy mined the waters between the Orkney Islands and Norway in an attempt to stop German U-boats from gaining access to the North Atlantic through the GIUK gap in the final phase of the war. For Russia, World War I at sea also provided painful lessons about just how constrained its access to the broader maritime domain was and just how vulnerable the access points it did have turned out to be. The Russian Baltic Sea Fleet saw itself bottled up and under pressure from Germany throughout the war, while the Black Sea Fleet was blocked from entering the Mediterranean and sailing further by the Ottoman Empire, which controlled access to and from the Black Sea through the Bosporus. Indeed, one of Britain's largest failures during the war, the Gallipoli campaign, was an attempt to open up access to the Mediterranean from the Black Sea for the Russian fleet there.

But World War I also gave Russia's leaders a first glimpse at the maritime possibilities inherent in its Kola Peninsula in northwestern Russia and also its potential vulnerability. The Kola Peninsula offered sheltered locations for ports where Russian ships would have relatively unfettered access to the Atlantic through the Barents and Norwegian Seas. And the Gulf Stream meant that the waters off the Kola Peninsula remained essentially ice free throughout the year.

The city of Murmansk was founded in 1916, in the middle of the Great War, and its location on the Kola Peninsula provided shelter and easy access to the Barents Sea for both merchants and warships. But the other combatants during World War I also realized the potential importance of the new Russian access to the North Atlantic via the Barents Sea. After Russia left the war in the wake of the Russian Revolution, Britain dispatched an expeditionary force to take control of the vital Kola Peninsula. The British force was delivered to the shores of the Kola Peninsula by twenty Royal Navy ships in June 1918 and was soon joined by additional forces from the United States. The decision by London to dispatch this force was in no small part driven by the need to stop Germany from gaining access to the resources in the region and to prevent Germany from seizing a new basing area for its submarines that would provide access to the North Atlantic from points not constricted by British submarine barriers, at a time when American reinforcements to Europe were in full swing.[11] The Murmansk expedition, which departed Russia in 1920 after a period of supporting the White side in the Russian civil war against the communists led by

Vladimir Lenin, is barely remembered in the West today. The joint British-American expedition would, however, leave more of a mark in Soviet and Russian strategic culture and would have implications for Moscow's thinking about the Kola Peninsula and its role in Russia's maritime strategy.[12]

CHAPTER 4

World War II

The Battle Expands

Interwar Developments

THE TWENTY YEARS BETWEEN the two world wars was not devoid of thinking on the potential for access to and control of the broader North Atlantic in coming contests. Indeed, both the British and Germans innovated based on their experiences at sea during World War I. But the conclusions drawn from the competition in the North Atlantic and adjacent seas were quite different. British naval authorities assumed that the next contest would proceed along the same trajectory as the one before, a common error in thinking often found on the side of the victor. Thus the British focused on enhancing the already existing tools and tactics that would help them win the next competition as well. The British, for example, invested significantly in the development of sonar technologies, a system that was first introduced late in World War I but was at that time too rudimentary and was based on passive detection of submarines, which rendered it useless except in the most ideal of conditions.[1] The Germans, on the other hand, focused on operational and tactical innovation and developed a doctrine that emphasized attacks against shipping at night while on the surface and using groups of submarines that could engage the targets in a convoy from multiple directions at once.[2] Germany was of course restricted in its development of the submarine force, with the Treaty of Versailles signed following the end of World War I explicitly forbidding Germany from maintaining a submarine force. Still, development of new submarines continued informally through design bureaus outside of Germany in an effort to maintain seed corn for future naval development; operational and technological advances accelerated drastically after 1935 when Germany was once again allowed to stand up a submarine force that could grow to a little less than half the size of the British submarine force. It was a case of doing the most with the limited resources available to Germany. Indeed, these constraints on German naval power drove Germany to focus heavily on its submarines because they were, after all, the naval weapon already proven to be able to offset the advantages of a superior surface force.

Strategic Setting

World War II began in Europe in the east in 1939 with the German invasion of Poland, but it soon expanded to the west with the German invasion of Holland, Belgium, Denmark, Norway, and France in 1940, with Germany's rapid advance also causing the ejection of British forces from the continent and back to the British Isles. Later, the war reached truly monumental proportions with Operation Barbarossa, Germany's invasion of Russia in 1941. The contest in Europe was of course centered on the domination of the European heartland, but the North Atlantic again played an important role as the highway for U.S. lend-lease shipments to both the United Kingdom and the Soviet Union, and later for the transfer of American forces across the Atlantic to Britain as the buildup began for the Allied landing in France in 1944. But already in 1942 the United States launched an ambitious transoceanic amphibious operation across the North Atlantic, when Task Force 34 transported nearly 34,000 American soldiers and their equipment from Hampton Roads to Morocco for Operation Torch, the British-American invasion of French North Africa.[3]

Here a fundamental strategic dynamic in modern European war was clearly reconfirmed: a war in Europe could not be won by controlling the North Atlantic, but the same war could certainly be lost there. This fundamental point was clear in the minds of the Allied political leadership. Winston Churchill declared after the war that "the Battle of the Atlantic was the dominating factor all through the war. Never for one moment could we forget that everything happening elsewhere, on land, at sea or in the air depended ultimately on its outcome."[4] The importance of access to the North Atlantic in keeping Britain in the war cannot be overestimated. Many of the absolutely vital supplies needed by Britain came across the North Atlantic from North America, including some 80 percent of the fuel needed by the United Kingdom.

The battle for control of the North Atlantic during World War II was far more intense and expansive than the contest at sea during the Great War. Indeed, the war in the North Atlantic was waged essentially from the first few days of the war until the bitter end in the late spring of 1945, making it the longest enduring campaign of World War II. Losses on both sides, by most measures, were also far higher during the second battle for the North Atlantic. As in World War I, ASW efforts during World War II proved to be resource-intensive. The Royal Navy alone fielded more than eight hundred escort ships, some forty escort carriers, and three hundred ASW aircraft against the German submarines.[5] And if ASW and modern naval conflict involving submarines was born in the North Atlantic during World War I, it made a significant leap forward in terms of sophistication in the same maritime domain during World War II.

Radar and Radios

While passive and active sonar are mainstay sensors for ASW today, they both played a limited role during World War II. Both the British and the Americans, though, launched ambitious development efforts in both passive and active sonar during the war. Toward the end of the war the British ASDIC system had progressed far enough to be able to track a submarine that had been detected using another type of system, such as radar or radio intercepts. But no matter the development in sonar, radar and radio intercepts provided the bulk of the sensing used by both sides during the battle of the Atlantic during World War II. Radar was effective in detecting German U-boats since they still had to spend considerable time on the surface, either to recharge their batteries or to move on the surface to conduct night attacks against convoys.[6] Radar could of course also be used by the U-boats to detect and track surface ships, but the Allies were at a distinct advantage in terms of range and capacity through much of the war.

Radio communications were also extensively used by both sides in the fight for control over the North Atlantic. This gave rise to radio interception and triangulation as methods to locate the source of the radio emissions. Radio interceptions were used by the Allies and the Germans to gain an understanding of where the convoys and the submarines were likely to be operating. This meant that both sides in the battle for control of the Atlantic placed a huge emphasis on the ability to quickly and accurately crack the opponents' encrypted radio communications. The British success in capturing an intact encryption machine, dubbed the Enigma, off the stricken but still floating *U-110* off the coast of Iceland in May 1941, and its subsequent exploitation for signals intelligence in the battle against Germany's submarines, may be the one sequence of events from the struggle at sea that remains in broader popular memory. It has even inspired Hollywood movies, of admittedly dubious accuracy.

U-110 had entered into the German fleet in late November of 1940. She previously had operated out of Kiel and on her first patrol was only a moderate success. During this first patrol *U-110* managed to damage only two merchantmen south of Iceland; the submarine was then forced to head for home for repairs to its damaged deck gun, a weapon very much in use during the early part of World War II. Later *U-110* was transferred to the German submarine base in Lorient in France, which offered the chance to operate in the North Atlantic without having to pass the relatively small and heavily patrolled North Sea and crossing the GIUK gap. On April 15, 1941, *U-110* left

Lorient for her second patrol in the North Atlantic. Early in the patrol *U-110* scored a victory by sinking a merchantman off the west coast of Ireland on April 27, and then headed north to seek out new targets between Greenland and Iceland. Here *U-110* found a Britain-bound convoy and managed to sink two additional merchantmen in short order. The British escort, however, aggressively responded and a corvette and destroyer pursued *U-110* and dropped depth charges, damaging *U-110* and forcing her to the surface. With the damaged *U-110* on the surface and the crew preparing to abandon the boat, both the British escorts and the German crewmen realized that *U-110* was not about to sink after all. The British quickly put together a boarding party to seize sensitive equipment on board *U-110*, including the Enigma machine and the associated code books. *U-110* was taken in tow back to Britain, but ultimately sank while en route. But the capture of the Enigma machine and the code books constituted a real coup. The boarding of a German submarine was a rare event during the war, and the fact that the German crew members were either dead or taken as prisoners of war meant that the German navy was unaware that its sensitive communications and cipher gear had been captured.

Contrary to popular belief, the capture of the German Enigma machine off *U-110* and the ability to decipher German radio communications did not provide the British and the Americans with what would be called "information dominance" in today's parlance. The signals intelligence gathered from radio transmissions covered only the orders and guidance provided from ashore to German submarines at sea and the radio notifications from submarines as they passed through the GIUK gap and moved into the broader North Atlantic. The Allies were still left guessing about the instructions that the German submarine commanders put to sea with, and Enigma certainly did not provide any insights about the deliberations and larger operational decisions made at the German submarine command. Furthermore, the German navy updated its ciphers on a monthly basis, which meant that the British code crackers had to start from square one after each change.

This caused a delay between signals interception and the creation of actionable intelligence. This gap was sometimes less than a day, but on other occasions extended to several weeks, plenty of time for operations and events to develop at sea. While the exploitation of the Enigma was perhaps not the decisive turn in the battle of the Atlantic that it sometimes is made out to be—no single technological factor rarely is pivotal in war—it did provide real intelligence value and allowed the Allies to reroute convoys around areas where German submarines had taken up station to intercept them. The intelligence gathered, however, was not specific enough for Allied ASW forces to detect, track, and successfully attack German submarines out in the Atlantic.

But breaking encryptions to glean intelligence about movements at sea was far from a one-sided affair. Germany successfully broke British naval codes in late 1941, which meant that the German submarine command had a better understanding of convoy movements in the North Atlantic throughout much of 1942 and could therefore more effectively position its submarines in locations and patterns that would make it easier to intercept the convoys.

The contest in the North Atlantic during World War II was in reality not decided by the Allies' ability to sink German submarines, although the German navy did lose a staggering number of them throughout the war. Instead, just as often the Allied ASW and escort forces were successful in detecting and engaging the German submarines so that they were forced to break off their attacks and depart the area, which points to an inherent and enduring vulnerability of any submarine: once detected and engaged there is little a submarine can do to defend itself against forces on the surface or in the air. This is a vulnerability that remains to this day.

The Battle for the Atlantic Ashore

The second battle for the Atlantic was far more expansive than the one fought during World War I. Operationally the struggle between German submarines and the Allied convoys and their escorts ranged across nearly all of the North Atlantic, from off the coast of Iceland in the north and down to the western tip of the African continent, and across to North America and into the Caribbean. Countries and regions in northern Europe that had been able to remain on the sidelines of World War I were pulled into World War II, in no small part due to their geography, which enabled extended operations in the North Atlantic. Better access to the Atlantic drove German motivations for occupying Norway. During World War I the German navy struggled to operate much beyond the North Sea, but submarine bases on Norway's coast would provide direct access to the broader Atlantic and could contribute to the siege laid against Britain after France fell in 1940. In addition, Germany was eager to cut off supplies and raw materials for Britain coming from the Baltic Sea region, which often were transported by rail through Sweden to Norwegian ports with access to the Atlantic. Seizing Norway and the ports there meant that Britain's supplies from the Baltic Sea countries could no longer pass through waters controlled by Germany.[7] The German push into France in 1940 was of course driven by nonmaritime grand strategic considerations, but it also provided Germany with additional submarine bases that allowed German submarines to range across the North Atlantic south of the GIUK gap and avoid the natural choke points there and in the North Sea. All in all, Germany established two submarine bases in Norway and another five in France. The bases included expansive hardened shelters and docks to make it far harder for Allied airpower to destroy submarines at pier. Some of these hardened submarine pens in Norway and France remain standing to this day.

Due to the far-ranging nature of the contest in the North Atlantic during World War II and the use of new technologies and platforms, not least long-range aviation, geographical locations away from the European continent that had been of little interest in the previous war came into play and had real strategic importance. Islands in the North Atlantic that had been of little value before now became veritable unsinkable aircraft carriers that could be used to launch air operations over the North Atlantic. The United States and the United Kingdom developed long-range aircraft capable of carrying significant loads, such as the PBY4-1, which could provide overhead cover over the large open ocean between North America and the GIUK gap. The Germans, meanwhile, used the Ju-87 Stukas for attacks against shipping and warships in coastal waters, including in the Norwegian Sea, the Mediterranean region, and the Barents Sea.

Iceland had been essentially untouched by World War I and was in any case neutral in the conflict, as until 1918 it was part of Denmark, a nation that managed to stay out of the Great War. But both sides in World War II quickly realized the strategic importance of Iceland as an overwatch position on the GIUK gap. British forces occupied Iceland in 1940 to deny the island to Germany, and U.S. forces followed in early 1942 and remained until the end of the war. The long-term importance of Iceland, even beyond the world war, as a toehold that enabled both air and maritime operations in the North Atlantic was not lost on Washington and London. A 1944 article in the influential American magazine *Foreign Affairs* authored by the prominent geostrategist Hans Weigert stated that "Iceland is a vital link in the relations of North America and Eurasia. From her shores much of the North Atlantic can be controlled. . . . This war has taught the British and ourselves that Iceland must remain forever an integral part of the mutual defense system of the two countries and must never be permitted to fall into the hands or under the influence of a potential enemy."[8]

Other specks of dry land north of the GIUK gap also played roles in the battle for the North Atlantic, albeit not in such a prominent fashion as Iceland. The Faroe Islands, which lie between Scotland and Iceland, were occupied by the British after the fall of Norway to ensure that the Germans did not get yet another toehold in the maritime domain from which to launch operations. Svalbard, north of Norway, was the object of both British and German operations and at times served as the basing area for German weather stations, with the resultant weather reports used to plan air and sea operations in the Barents Sea against the Allied convoys to Russia.[9] In the southern part of the North Atlantic the Azores, roughly eight hundred nautical miles west of the Portuguese coast, caught the eye of both

Washington and London as a potential base for airborne ASW units to help cover the North Atlantic south of the GIUK gap, an area that until then had been relatively weakly patrolled by the British and Americans. Neutral Portugal was at first reluctant to grant the Allies basing rights, fearing that this would trigger a German attack on Portugal. But with the Allied advances in North Africa that risk receded considerably, and the Azores were opened up for Allied access in 1943.

The War in the High North

But the western part of the North Atlantic, and the routes from North America to Britain, were not the only crucial sea space under contest during World War II. The eastern North Atlantic, along with the Norwegian Sea and the Barents Sea, soon became vital for the Allies in keeping the Soviet Union in the war. After Germany's invasion of the Soviet Union, assistance was extended by the British and the United States to supply arms, vehicles, fuel, and ammunition to the beleaguered and poorly supplied Soviet forces trying to fend off the German advance. The United States had three viable options for supplying the Soviet Union with aid: across the Pacific, through the Middle East and into Russia's south, or across the Atlantic and through the Norwegian and Barents Seas for landing at Soviet ports on the Kola Peninsula. Due to geography and the presence of Axis navies in the Mediterranean, only the last option was available to the British. The effort to supply the struggling Soviet Union with war supplies was rapidly put together in the wake of the German invasion of the Soviet Union in late June 1941. By August the same year, the first small convoy of Allied ships left port in Iceland and headed toward Russia across the eastern North Atlantic and the seas north. Iceland remained a crucial node throughout the convoy effort, as American and British supply ships gathered there to form the joint convoys for the journey lasting roughly ten days over the sea and to the Kola Peninsula and beyond. Between 1941 and the end of the war in Europe, some nine hundred merchant ships, along with escorts of destroyers and corvettes, made their way into the far North Atlantic to resupply the Soviet Union. The weather conditions were often harsh this far north, and they were made worse by the Arctic ice that many of the convoys came into contact with as they sought to stay as far away as possible from German submarines, surface ships, and aircraft based out of northern Norway. The convoys across the far North Atlantic delivered some four million tons of supplies to the Soviet Union, but more than 10 percent of the shipping was sunk by German submarines, surface ships, and aircraft. The convoy escorts destroyed some thirty German submarines in return during the convoy campaign, but the escort job was made more difficult because real cooperation between the U.S. and Royal navies on the one hand and the Soviet navy on the other never really materialized, and the Soviet navy proved unable to provide much protection during the final transit across the Barents Sea, when the convoys were at their most vulnerable due to the proximity to northern Norway.

The Second Battle of the Atlantic Draws to a Close

Germany continued to innovate in the field of submarine technology until nearly the end of the war, and it introduced several technologies that could, under different circumstances, have helped turn the contest in the North Atlantic. These included more powerful radar detectors as well as the snorkel, which allowed the submarine to recharge its batteries while submerged. This considerably reduced the Allies' ability to detect Germany's submarines using radar. Ultimately, however, the Allies were able to hold off the more sophisticated German submarines until the Third Reich crumbled under the combined onslaught of Allied ground forces moving in on Germany from both the west and the east.[10]

The ultimate German failure in its campaign to close off the North Atlantic as a reinforcement and supply bridge between the Allies of Europe and North America lay not only in its inability to overcome the awesome production capacity of the United States, which built new supply ships, other war materiel, and other consumables at a dizzying rate. During one period U.S. shipyards completed three Liberty ships, the standard class of cargo ship during the war, on a daily basis.[11] Another important and mostly underappreciated factor was the German failure to distinguish between and prioritize among the most important targets found at sea. The German submarine command made no distinction between ships that carried troops, fuel, and tanks, or beans, lumber, and steel. No U.S. troop ships coming across the North Atlantic were sunk until the period of the Battle of the Bulge, around Christmas of 1944, at a time when Allied ground and air power were well established on the European continent. This is not to say that halting the supply of the materiel that kept Britain as a society and economy in the war was unimportant. But a clearer focus on halting the buildup and sustainment of ground and air power across the sea could have greatly complicated the Allies' plan for the return to the European continent.[12]

CHAPTER 5

The Cold War

The Long Third Battle for the North Atlantic

WHILE TWO DECADES SEPARATED the first two battles of the Atlantic, the third contest opened nearly immediately after the end of World War II as the erstwhile allies of that conflict began their competition over the future of Europe and the global system more broadly.[1] But the third battle of the Atlantic would be very different from the first two. While the two previous contests had been sharp and violent and relatively short, the competition in the North Atlantic during the Cold War would last for more than four decades, with essentially no shots fired. This, however, did not mean that the third Atlantic contest was without its moments of tensions and real danger. In terms of resources spent on ships, submarines, aircraft, bases, sensor networks, and personnel, it would be the most expensive and expansive of the Atlantic competitions yet. But unlike the first two contests in the Atlantic, the Western allies held much of the early advantage in the competition. But the Soviet Union slowly but surely gained on the United States and NATO in terms of capabilities and quiet submarines as the Cold War progressed. The third battle would also further contribute to technological and strategic developments that were arguably as important as those achieved during the two world wars; many of these are with us to this day in the North Atlantic and in other important maritime domains.

The Cold War is of course best remembered by the arraying of NATO and Soviet ground forces in Central Europe and the potential of World War III as essentially a clash of great armies on the plains of Germany and of air forces above, which could likely develop into a nuclear exchange between the two sides at some point. But those images in public memory underappreciate the role of the North Atlantic under any crisis scenario between the Soviet Union on the one hand and the United States and NATO on the other. Just as during the two world wars, the Atlantic provided the lifeline between North America and Europe, and the ability to ship reinforcements across the sea in the case of a war remained central to American defense planning for Europe. Indeed, that NATO, the alliance founded to defend Europe against Soviet aggression, has North Atlantic in its name says quite a bit about the centrality of the North Atlantic in the Cold War and the maritime space that the great Western alliance formed around. But the introduction of nuclear weapons also meant that the North Atlantic would play another role in the new contest over the future of Europe. Over the decades, the Atlantic would become the home of key parts of both sides' nuclear second-strike capability, which would bring new dangers to the competition at sea.

The Soviet Navy and the North Atlantic

The Soviet Union emerging victorious from World War II had little in the way of naval power, even though the Soviet navy had had one of the largest submarine forces in the world during the early part of the war.[2] But that began to change nearly as soon as it was clear that the world was moving to a new contest between the West and the Soviet bloc. The Soviet Union's own war experience in Europe during World War II was predominantly a ground affair, with the great clashes of the war's eastern front such as Stalingrad, Kursk, and the Battle for Berlin in fresh and fervent memory. This, of course, stands in stark contrast to the American experience of amphibious landings in France, Italy, and North Africa, along with the campaign in the Pacific against Imperial Japan, an effort characterized by island hopping and several significant naval battles. The Soviet navy, meanwhile, had put in a modest effort in coastal areas during the war with only limited effect at the end of the day, including in the Atlantic approaches to Soviet ports in the Barents Sea. Indeed, Soviet sailors and marines had frequently been taken from their ships and bases and thrown into ground operations during the war to supplement badly battered army units. But Moscow had been impressed with the Western Allies' ability to project force across the oceans during World War II, both in the North Atlantic and in the Pacific, and the use of the sea-lanes to sustain far-reaching military operations with supplies, equipment, and reinforcements.[3] Also, the memory of incessant German attacks from northern Norway against shipping and warships in the Barents Sea was clear. In addition, the vulnerability of the Soviet Union's access to the sea during the war, where the Soviet Baltic and Black Sea Fleets had, just as during the Great War, remained bottled up for much of the war, was also clearly on Soviet minds. As the Cold War progressed, this Soviet view of the Western use of sea power was further reinforced by the Korean and Vietnam wars, where Soviet naval leaders could see America projecting force onto land from the sea and conducting air strikes from aircraft carriers with virtually no challenge from the opponent.[4]

From a Soviet perspective then, the sea, and in particular the North Atlantic, would play an important supporting role in any future new conflict with NATO and the United States in Central Europe. By 1961, a mere fifteen years after the start of the Cold War, the Soviet Union had expanded its navy considerably, adding more than four hundred submarines. And while the Soviet Union also added major surface combatants, such as frigates and destroyers, the enduring and growing Soviet emphasis on subsurface warfare to counter NATO's maritime power in the North Atlantic and elsewhere was clear from the beginning. In 1961 one-fifth of the Soviet Union's overall naval tonnage was made up of submarines. This proportion grew to nearly half of the overall tonnage by 1981.[5] The Soviet Union introduced its first nuclear submarine in 1958 and then rapidly filled its force with *November*- and *Echo*-class nuclear attack submarines (SSN). In the decade following the introduction of nuclear propulsion, the Soviet Union built some fifty-five nuclear submarines. These new classes of SSNs could evade radar detection (since there was no need to surface while on patrol) and came with greatly extended operational range, but they did generate significant noise through water flows across the hull and loud cavitation from the propeller.[6]

The scope and reach of the Soviet Union's naval power grew rapidly under the legendary leadership of Admiral Sergey Gorshkov, who served as the commander in chief of the Soviet navy from 1956 to 1985. Having joined the navy in 1927, Gorshkov brought the Soviet navy into the nuclear age and transitioned it from largely a coastal force that had proved so inadequate during both world wars to a navy with global reach and the capacity to challenge NATO's dominance in the North Atlantic. Along with submarines, the Soviet Union operated larger and more powerful warships, including destroyers, nuclear cruisers, and, toward the end of the Cold War, a modest number of aircraft carriers. The growing Soviet fleets were also given more expansive peacetime missions, including naval diplomacy in far-flung places such as Latin America, the Middle East, and Africa, with many of these missions being supported by the Northern Fleet and through the Atlantic. But even with these more expansive naval missions around the world, the core wartime missions of the Soviet navy remained denying the maritime approaches to Russia and stopping NATO's use of the sea-lanes for reinforcements.

Like the United States, the Soviet Union sought to use the maritime domain for part of its nuclear deterrent. The Soviet Union built a huge number of SSBNs starting in 1958 that would patrol in the Atlantic and the Pacific. As the ranges of the ballistic missiles grew, the Soviet SSBNs were able to patrol closer and closer to home waters, which were less vulnerable to detection and attack by U.S. submarines. This ultimately led to the Soviet bastion system around the Kola Peninsula, where Soviet SSBNs could remain in the Barents

Sea, protected by Soviet attack submarines and surface ASW forces, making the Soviet nuclear deterrent more defendable against potential U.S. surprise attacks. The bastion system further increased the importance of the Kola Peninsula to the Soviet Union's (and later Russia's) defense strategy.

While denying NATO access to the North Atlantic to keep naval forces away from the Soviet Union was important as a matter of course, the growing reach of the Soviet navy into the North Atlantic was also driven by the increasing range of the air power on board American aircraft carriers, which could play both a conventional and nuclear role during a war. In the late parts of World War II, U.S. carrier aviation had achieved a range of nearly 800 nautical miles while carrying relatively modest payloads. By the late Cold War this range had been extended to more than 1,500 nautical miles by, among other things, using in-air refueling. The carrier aircraft could also carry considerably heavier payloads that now could include nuclear weapons.[7] This led to a Soviet naval strategy of extending the Soviet sea denial zone by meeting U.S. and NATO naval forces further out in the North Atlantic, including south of the GIUK gap.

The role of the air domain in the competition over control of the North Atlantic, which first became important during World War II, continued to grow during the Cold War. For example, the TU-16 Badger and TU-22M Backfire, both long-range bombers first introduced in 1954 and 1962, respectively, were operated in large numbers (some 350) by Soviet naval aviation and were equipped with long-range antiship missiles. These were intended to leave their bases on the Kola Peninsula, enter the North Atlantic region, and engage U.S. naval forces and shipping with large barrages of missiles that could overcome U.S. shipboard air defenses. At its height, Soviet naval aviation was one of the largest air forces in the world and included some 1,400 aircraft ranging from fighters to long-range strike aircraft, MPAs, and reconnaissance planes.[8] The Soviet use of air power in conjunction with submarines to attack shipping and warships was in no small part informed by a lesson learned from

World War II. The Soviets had concluded that the German U-boat campaign and use of air power against shipping failed because it focused exclusively on transport shipping and was never combined to directly threaten the Allies' ASW forces.[9]

While attacks against ships with aircraft armed with antiship missiles were never attempted during the Cold War, the British experience during the Falklands War suggests that it would have been an effective Soviet method indeed. During that conflict the Argentinians were highly successful in using antiship missiles delivered by aircraft against Royal Navy surface ships, and they leveled a devastating attack against, among others, the SS *Atlantic Conveyor*, a commercial Roll-on/Roll-off ship.[10] The *Atlantic Conveyor* had been requisitioned by the Royal Navy to carry vital supplies and equipment for the British landing force, and it had on board several helicopters intended for tactical lift during the British landing and recapture of Port Stanley, the capital of the Falklands Islands. On May 25, 1982, the *Atlantic Conveyor* was hit by two French-made Exocet antiship missiles fired by Argentinian attack jets. Most of the helicopters were destroyed in the resulting fire, and the *Atlantic Conveyor* sank three days later while under tow. The British landing force then had to road march to Port Stanley rather than being air assaulted into the city, a significant additional risk to the operation. It proved a bitter lesson in the vulnerability of reinforcement shipping to attacks at sea.[11]

As the Cold War at sea progressed, the Soviet navy focused increasingly on adding cruise missiles to its submarines and surface ships, for attacks against both land targets and ships. This added considerably more range to the striking power of Soviet submarines and ships and increased the ability to strike against both cross-Atlantic shipping and American carrier groups operating in northern waters. Toward the end of the Cold War the Soviet Union also fielded new classes of nuclear guided missile submarines (SSGNs), the *Oscar-I* and *Oscar-II*, which could carry the long-range antiship cruise missile P-700 Granit, which had a range of nearly 400 nautical miles and a 1,500-pound warhead. The concept was that the SSGNs would attack Western shipping in concert with the Badgers and Backfires in the air, presenting the naval force with an over-whelming number of cruise missiles that the air defense systems would be unable to deal with.[12]

By the 1970s the Northern Fleet had increased its days at sea in the North Atlantic by almost 400 percent in comparison to the early 1960s,[13] and the trend continued into the 1980s. During that period, the Soviet navy went from having its warships and submarines spend some 1,600 days at sea in the Atlantic in the first year of the 1960s to more than 18,000 days in the mid-1980s, or some 30 percent of the overall share of Soviet navy days at sea around the world.[14] During the same period, Soviet submarine activity in and around the GIUK gap increased by some 60 percent in comparison to the early Cold War.[15] The Soviet Union also put its growing sea power on display for the United States and NATO during big naval exercises in the North Atlantic and elsewhere. During the exercise Okean (Ocean) in 1970, some sixty ships and submarines from the Northern Fleet par-ticipated in a force of two hundred ships from all the Soviet Union's fleets. At the time, this was the largest naval exercise conducted by any navy since World War II. As part of this exercise the Northern Fleet gathered ten surface warships, some thirty submarines, and several hundred aircraft for an attack against an enemy approaching Europe through the GIUK gap. This concept of sea denial opera-tions was further refined during the late part of the Cold War, to a point that it was assumed that the Soviet Union's extended sea denial effort could reach beyond the GIUK gap.[16]

While Soviet sea power grew in both scope and sophistication throughout the Cold War, it is important to remember that it was never an omnipotent force. Crew training was at times rudimen-tary, and morale among the ranks was low, even when compared to some of the personnel troubles the U.S. Navy experienced during the Vietnam War era. Proper maintenance was an enduring problem for the Soviet navy, and it drove close Western observers (such as naval attaches or others visiting Soviet ships) to question whether some weapon systems were actually functional. Still, the sheer mass of ships, submarines, and naval aviation that the Soviet Union generated posed a real challenge for the U.S. Navy and NATO in the Atlantic, especially considering that the Soviet navy would play a sea denial, rather than sea control, role during hostilities.[17]

The Most Militarized Region in the World

The rise of Soviet sea power in general and in the North Atlantic in particular also gave the Kola Peninsula crucial importance as a center of gravity for a Russia with global naval reach. Given the Soviet Union's complicated maritime geography with constrained access to the global maritime domain, the Kola Peninsula and the surrounding maritime domain out to the North Atlantic served four strategic military purposes for the Soviet Union. First, it was the ideal basing area for the Soviet submarine-based nuclear deterrent, and some 60 percent of the Soviet SSBNs were based there, surrounded by the aforementioned bastion of SSNs, surface warships, and potent air defenses. Second, the Kola Peninsula offered easy access to the Norwegian Sea, where in the late Cold War it was expected that U.S. naval forces would launch air campaigns against the Soviet Union with both conventional and nuclear munitions, as part of the more forward-leaning U.S. maritime strategy. Third, the northwestern Arctic around the Kola Peninsula offered an excellent forward operating and refueling area for Soviet strategic bombers preparing to strike targets in Europe and North America. And finally, the region served as the jump-off point for Soviet naval forces sallying into the central Atlantic to attack U.S. shipments of reinforcements from the continental United States to Europe.[18]

Throughout the Cold War the Kola Peninsula experienced fast and extensive growth as the region developed into the Soviet Union's primary springboard into the North Atlantic and for extended global power projection. As late as the 1950s the Northern Fleet included only a few surface warships and a mere 10 percent of the Soviet submarine force. Twenty-five years later the Northern Fleet had developed into the most powerful and capable of the Soviet fleets and included half the boats in the growing Soviet submarine force.[19] The Kola Peninsula, which sits in Russia's northwest right next to Norway and Finland, juts out like a thumb from the Scandinavian peninsula. About the size of Virginia, the Kola Peninsula was for long an underdeveloped part of the Russian empire, with only a small population of fishermen and indigenous people eking out a living from hunting and fishing. This all changed in the late 1800s when the peninsula saw major development to extract coal and rare metals, such as nickel, that had been discovered there. The early Soviet Union sought to develop the peninsula further in its bid to industrialize and bring under control the vast Russian Arctic, which was seen as crucial to the ultimate economic success of the Soviet system. Later the Kola Peninsula became an important part of the Soviet gulag system, where many prisoners were sent to provide hard labor for the Soviet Union, in the copper, nickel, and iron mines in the region.

The serious development of the Kola Peninsula as a military hub for Russia started only in the twentieth century. A rail line to the Kola coast from Saint Petersburg on the Baltic Sea was completed in 1915, and ports were expanded as part of the Russian war effort during World War I. The young Soviet Union sought to further connect the Kola Peninsula to the rest of Russia by building a canal between the Baltic Sea and the White Sea, which would require some 100,000 of the already mentioned gulag laborers and would open in 1933.[20] Tsarist Russia maintained a small squadron of ships on the peninsula for its Arctic Ocean Fleet beginning in 1916, but the squadrons were smashed and scattered during the Russian civil war. The origins of the modern Northern Fleet can instead be found in 1933, when six surface ships and three submarines were ordered to the peninsula through the recently finished White Sea–Baltic Sea canal. With the growth of the Northern Fleet during the Cold War came further development on land with infrastructure and industries associated with the navy, and a growing population. Between 1960 and 1990 the population of the Kola Peninsula nearly doubled from around 600,000 to close to 1.2 million, significant in itself, but especially impressive considering that only about 15,000 people lived in the region in 1920.[21] The Kola region also became, in relative terms, the most urbanized region north of the Arctic Circle anywhere in the world.

The Kola Peninsula's coastal geography made it an excellent location for Soviet naval bases. The Kola Peninsula's Barents Sea coast is dotted with a number of deep bays and fjords that proved to be well-protected and dispersed locations for the Northern Fleet's ship facilities. And the area around the White Sea just south of the Kola Peninsula offered further protected areas for the Soviet Union's naval forces. During the Cold War the Northern Fleet's ships and submarines were based at no fewer than eight facilities strung out along the northern coast of the Kola Peninsula, with the headquarters of the Northern Fleet at Severomorsk, a city that is part of the larger Murmansk metropolitan area. All in all, the Northern Fleet included more than two hundred vessels in the late Cold War period. The ships, submarines, and their facilities were further complemented by sixteen air bases and nearly three divisions of ground forces, including naval infantry and maritime special operations forces, as well as powerful air defense networks.[22] These naval bases were further complemented by facilities for ship and submarine repair and maintenance, nuclear waste storage, munitions stockpiles, and long-term storage of decommissioned ships and submarines.[23] During the Cold War, Western intelligence agencies estimated that some 50,000 people were employed in the naval industrial base in northwest Russia.[24] Severodvinsk, which sits across from the Kola Peninsula on the southern coast of the White Sea, served as the home of the Soviet Union's main naval shipyard, which, among other things, built the Soviet Union's first nuclear-powered submarine. During the Cold War it constructed some 120 nuclear submarines, or nearly half of the Soviet's total.[25] Indeed, the submarine yard in Severodvinsk grew into the largest submarine construction and maintenance facility in the world, complete with three covered and heated construction halls. The growth of Shipyard 402, the formal name for the shipyard facilities at Severodvinsk, caused Adm. Hyman Rickover, the legendary father of the American nuclear submarine force, to quip that it had "several times the area and facilities of all of the US submarine yards combined."[26] Taken as a whole, the naval region of the Kola Peninsula is not unlike that of the Hampton Roads area of Virginia, which also has a dense concentration of American naval power, in combination with the yards and other facilities to service and support that naval power.

As a result of the Soviet naval buildup, by the mid-1970s there was a growing sense inside the U.S. Navy that the Soviet Union was catching up in terms of sophistication and capacity and that the North Atlantic would be under serious threat of being closed to American reinforcements if there were a conflict in Europe.[27] In addition, NATO's northern flank with Denmark and Norway was seen as extremely vulnerable and at risk of being left behind the screen of the powerful Northern Fleet and under threat of land invasion from the Kola Peninsula. The concerns about the growing power of the Soviet navy in the Atlantic and NATO's exposed northern flank were further informed by the shrinking size of the U.S. Navy in the post-Vietnam period.[28]

NATO and the North Atlantic

The U.S. and NATO efforts to respond to the Soviet Union's growing sea power in the North Atlantic and elsewhere were no less comprehensive than Soviet efforts built on the recent experiences of World War II. The U.S. and NATO's combined efforts grew over time into a truly networked approach that combined platforms, sensors, and operations on the surface, in the air, and under the water. NATO's approach to defending the North Atlantic effectively leveraged the Western alliance's geographical advantages and the command of the choke points of the North Atlantic, namely northern Norway, Iceland, Greenland, and the United Kingdom. What emerged from this was a barrier strategy around the GIUK gap, where sensor networks, MPAs, ships, and submarines would be able to detect and attack Soviet submarines before they entered the broader North Atlantic south of the GIUK gap. As the Soviet Union added submarines to its nuclear deterrent, the barrier strategy also took on an important role in defending against Soviet nuclear strikes against the United States. The early Soviet SSBNs had to cross the North Atlantic and patrol close to the U.S. east coast to be able to launch its short-range nuclear missiles against targets in America.

With the growing sophistication of submarines on both sides of the Iron Curtain and the increasing ability of those submarines to remain submerged for longer periods of time, passive and active sonar became the mainstay method to detect, track, and target submarines. Along with the sonar systems and other sensors on board ships, aircraft, and submarines, in 1949 the United States began to develop a chain of passive sonar arrays called the Sound Surveillance System (SOSUS) for detecting and tracking Soviet submarines. At first the SOSUS network developed slowly, with stations and arrays established close to the U.S. east coast and off the Caribbean. But the initial results were remarkable, with the U.S. Navy being able to detect Soviet submarines in the Norwegian Sea from arrays on the other side of the North Atlantic. In its early years of operation SOSUS also played a role in detecting the four Soviet Foxtrot submarines from the Northern Fleet that left the Kola Peninsula and headed across the North Atlantic to Cuba during the Cuban missile crisis. And the SOSUS system grew tremendously in scope and size throughout the Cold War. In the mid-1970s the system consisted of twenty-two onshore facilities, of which thirteen were focused on the Atlantic, manned by some 3,500 personnel. SOSUS arrays were also installed farther north and east in the North Atlantic, and an onshore station was emplaced on Iceland in 1966. An additional network was also emplaced off the northern coast of Norway in 1963 in order to cover the northern entrance of the Norwegian Sea from the Barents.[29] By 1981 North Atlantic onshore SOSUS installations included locations in the United States, Canada, the United Kingdom, Bermuda, Norway, Iceland, the Azores, and Denmark.[30] In the 1980s the staffing grew to more than 4,000, while the number of onshore sites declined due to advances in technology that meant fewer stations could monitor a larger number of arrays in the oceans. By the late Cold War, SOSUS was integrated into the Integrated Undersea Surveillance System (IUSS) and also complemented by fourteen SURTASS ships, vessels with towed sonar arrays that could be dispatched to places where additional sonar coverage was needed; eight of these were dedicated to operations in the Atlantic and the adjacent seas.[31] Increasingly capable MPAs, such as the P-2 and P-3, were used in conjunction with SOSUS, where SOSUS tracks would be used to cue the MPAs to the likely location of Soviet submarines for further tracking.[32]

All in all, the United States spent some $16 billion on building, expanding, and maintaining the SOSUS network throughout the Cold War in the Atlantic and the Pacific. The enterprise included laying some 30,000 miles of cables along the ocean floor. The SOSUS effort was further supported by a flotilla of ships that could perform repairs on the cables. As late as 1991 the U.S. Navy maintained three of these ships.[33] The SOSUS system proved highly effective and one that the Soviet navy had difficulties fully understanding and countering. But the John Walker spy ring, which was broken up by the FBI in 1985, helped the Soviet Union to get a fuller understanding of SOSUS' capabilities and how and where to operate to reduce the chances of detection in the North Atlantic, as well as how to quiet future generations of Soviet submarines.

The West's military-technological response to the Soviet challenge in the North Atlantic was supported and underpinned by political-military efforts ashore that helped build and deepen the alliance around the Atlantic that would sustain U.S. and NATO efforts in the maritime domain throughout the Cold War. The effort shaped significant parts of NATO's Cold War command structure and planning, as well as the kinds of militaries and navies that were built and maintained by the NATO allies in the immediate North Atlantic area.

In 1952 the new transatlantic alliance created a structure with two supreme allied commanders: Supreme Allied Commander Europe (SACEUR) and Supreme Allied Commander Atlantic (SACLANT). Broadly speaking, SACEUR—always a U.S. four-star flag officer, the first being Gen. Dwight D. Eisenhower—would be responsible for the land and air domain in Europe, while SACLANT saw to operations in the broader North Atlantic and the potential challenge of getting U.S. shipping across the Atlantic in the teeth of Soviet naval interdiction. Allied Command Atlantic (ACLANT) was further divided into subregional areas of responsibility, such as northern

Atlantic (NORLANT), eastern Atlantic (EASTLANT), and western Atlantic (WESTLANT), where allied naval commanders would have responsibility for tasks such as ASW and ensuring the security of the sea-lanes for shipping and allied naval operations. NATO's ACLANT was responsible for a number of major naval exercises during the Cold War, such as the recurring training events Northern Wedding, Teamwork, and Ocean Safari, which took place in the North Atlantic, the North Sea, and the Norwegian Sea, to test the readiness of NATO member naval forces and their ability to work together to maintain control of the sea. For continual operations, a Standing Naval Forces Atlantic (STANAVFORLANT) was created, a construct where allied nations would contribute ships to the force on a rotating basis.

U.S. naval force structure was also adapted to be able to better respond to the third battle of the Atlantic during the Cold War. In 1950, shortly after the creation of NATO, the U.S. Second Fleet was stood up and headquartered in Norfolk, Virginia. Second Fleet was responsible for operations in the North Atlantic and to assure control of the sea lines of communication between North America and Europe during war. While engaged in exercises and preparing Atlantic-based U.S. naval units for deployments with other fleets throughout the Cold War, Second Fleet also saw its fair share of real-world operations in the Cold War struggle, albeit not in connection to its core mission of maintaining sea control against the Soviet navy. Among other things, Second Fleet was engaged in the naval blockade operations during the Cuban missile crisis in 1962, and it provided the amphibious elements of the invasion of Grenada in 1983.

The Cold War also meant that Europe's pre-eminent naval force, the British Royal Navy, would focus much of its energy on keeping the North Atlantic open in case of war with the Soviet Union.[34] The Cold War contest in the North Atlantic coincided with the end of the British Empire in the years after World War II. In the eighteenth and nineteenth centuries the Royal Navy had had global ambitions and had guarded the international sea-lanes and projected British might onto the European continent and beyond to Britain's colonies and holdings. But World War II had exhausted the empire, after

Britain had already expended itself to near the breaking point in World War I. This meant that as the reach of the British Empire shrank so did the ambitions of the Royal Navy. And as the Cold War went into full swing, the Royal Navy found itself with less of a global role, but with a central function as the defender of the maritime lines of communication in the northeast Atlantic, along with providing the United Kingdom with the primary part of the nation's nuclear deterrent on board submarines.[35] Specifically, along with the peacetime mission of playing a role in the monitoring of the North Atlantic and the GIUK gap, during war the Royal Navy was expected to operate well north of the GIUK gap to intercept the Northern Fleet and deploy amphibious forces to protect NATO's northern flank.[36]

Later, during the 1970s, West Germany's navy, the Bundesmarine, was also drawn into the North Sea and the North Atlantic. Since its reconstitution after the end of World War II it had primarily focused on the Baltic Sea in a sea denial role to halt the Soviet Baltic Sea Fleet from exiting the Baltic Sea through the Danish straits in wartime. This transition to a North Atlantic focus was in no small part driven by a rising German recognition of the growing power of the Soviet Northern Fleet on the Kola Peninsula and its implications for U.S. reinforcements across the sea to Europe.[37] Germany was also nudged by Washington to take up a piece of the responsibility in the North Atlantic, as the U.S. Navy now had to draw off some naval resources to patrol the Indian Ocean and the Persian Gulf in response to the Soviet invasion of Afghanistan and the Iran-Iraq War, which threatened the vital sea-lanes of communications in the Arabian Gulf.[38]

The small country of Norway played an outsized role on NATO's northern flank during the Cold War, given its unique location as the closest NATO territory to the Barents Sea and the Kola Peninsula. This gave NATO a front-row seat to the operations and exercises of the Northern Fleet in the Barents Sea, and it also meant that any Soviet advance into the North Atlantic would have to pass by Norway.[39] The Cold War Norwegian navy was singularly focused on operations in the Norwegian Sea, the Barents, and the North

Atlantic and, at the height of the Cold War, included some fifteen submarines and a relatively large number of frigates and torpedo boats focused on ASW and attacking Soviet surface ships with antiship missiles and torpedoes. Norway's geography also offered unique opportunities for infrastructure related to the defense of the North Atlantic. In 1952 Norway opened the Andoya airfield in northern Norway, which hosted Norwegian MPAs for patrols in the Barents Sea and the Norwegian Sea, and thereby provided an early opportunity to detect and track submarines from the Northern Fleet. Norway also constructed Olavsvern submarine base in northern Norway, which provided easy access to the Barents Sea. The base included submarine pens sunk into the mountainside for protection against air raids. The same geography, however, meant that Norwegian territory was potentially of interest to the Soviet Union in wartime in order to make the Northern Fleet's access to the North Atlantic easier. Partly due to this, the United States forward-based equipment, vehicles, and munitions for a Marine Expeditionary Brigade for the U.S. Marine Corps to rapidly come to Norway's aid in case war broke out.[40]

NATO as an alliance did not own any forces, ships, submarines, aircraft, or bases, nor does it today. Instead, NATO had to rely on contributions from its members; and in the contest for control of the North Atlantic, real estate could be as important as ships or submarines. As during World War II, Iceland was crucial to keeping the North Atlantic open during the long contest between NATO and the Soviet Union. As World War II came to an end Iceland briefly considered neutrality as a foreign policy option, but as the fragile peace after World War II set in and morphed into a Cold War between the Western and Eastern blocs, the government in Reykjavik quickly realized that a neutral posture could not be sustained given two urgent realities: first, Iceland had no defense force of its own, and second, a war between the West and the Soviet Union would quickly give rise to a race to occupy Iceland in order to control a broad swath of the North Atlantic. Iceland therefore threw its lot in with the growing transatlantic alliance and signed a bilateral defense agreement with the United States, which allowed the basing of U.S. forces in and American operations from Iceland.

The U.S. presence in Iceland grew substantially during the Cold War, and at its height the United States maintained some five thousand service members on the island. In addition, U.S. and NATO military facilities employed another one thousand Icelanders in various capacities. Maritime patrol was one of the key missions flown out of the Keflavik airbase, located roughly thirty miles from the Iceland capital of Reykjavik. But as the Soviet Union advanced its air capabilities, especially long-range bombers with antiship missiles, the U.S. presence in Iceland took on additional air defense and reconnaissance missions. By the early 1980s the Keflavik presence included not only U.S. MPAs, but also AWACs planes, F-15 fighters, tankers, and a search-and-rescue squadron. During the same period, the United States also added hardened airplane shelters (to protect against a surprise Soviet air attack on Iceland) and expanded fuel and ammunition underground storage that would allow for nearly forty-five days of combat operations without resupply from the outside. Along with facilities that directly supported U.S. air missions around Iceland, the island also hosted radar installations, a SOSUS shore station, and a NATO communications facility. The 1980s proved to be an especially busy time in the air and maritime domain around Iceland. Along with the frequent antisubmarine patrols out of Iceland, U.S. fighter jets intercepted some 170 flights in the North Atlantic by Soviet air force and naval aviation aircraft.

As the southern node of the GIUK gap, Scotland also rose in prominence during the Cold War as a base for British and allied maritime operations in the North Atlantic. The RAF base at Kinloss, which had served as a small pilot training center during World War II, became the hub for the British MPA fleet, which could easily cover the GIUK gap and range into the Norwegian Sea from northern Scotland. The Royal Navy base Clyde on the west coast of Scotland, often called Faslane by British and allied sailors, grew into the main basing area for the Royal Navy's surface fleet and also the British submarine-based nuclear deterrent. Between the early 1960s and the end of the Cold War the United States also maintained a forward base at nearby Holy Loch for its Atlantic SSBNs. The base was used for repairs and maintenance of submarines and as a location for crew swaps, thereby saving the SSBNs the long trip home to the U.S. east coast across the North Atlantic.

Other small pieces of solid land in the Atlantic also took on supporting roles during the Cold War as toeholds and outposts in the maritime vastness of the North Atlantic. The small Faroe Islands, a cluster of islands between Iceland and Scotland with a population of roughly 50,000, and which belongs to Denmark, had been occupied by a small British force during World War II, but the Cold War brought a more extended U.S. and NATO presence of sorts. The Faroe Islands were too small and remote to support a major basing of forces, but the United States installed Loran C, a radio navigation system used by warships, and early warning systems that covered parts of the eastern North Atlantic. The Faroe Islands have always been uneasy about their linkage to Denmark, and moves have been made toward independence at various times. Therefore, during the Cold War Washington nudged Denmark to increase its economic subsidies to the Faroe Islands to ensure that Denmark, and by extension the United States and NATO, did not lose access to this piece of land in the middle of the GIUK gap.[41] Similarly, and a little further south, the Shetland Islands, which are part of Scotland, played host to a major air defense radar during the Cold War at RAF Saxa Vord, with the mission to detect and track Soviet naval aviation forays into the North Atlantic.

Meanwhile, the Azores, an archipelago in the mid-Atlantic belonging to Portugal, became an important base of operations for covering the mid-Atlantic gap, south of the GIUK gap, with U.S. MPA patrols. Throughout the Cold War the United States operated P-2s, and later P-3s, from the Azores. First constructed during the late stages of World War II, the U.S. Lajes air base there was further expanded in the 1980s in response to the widening reach of the Soviet navy. The Azores also served as an important refueling point for U.S. military flights to and from Europe and as a staging base for U.S.

tanker aircraft. Toward the end of the Cold War some 250 U.S. and allied military flights went through the Azores on a monthly basis.[42]

The network of bases and sensors that were established throughout the North Atlantic region and beyond to deal with the challenge of the Soviet navy was truly impressive, and by the end of the Cold War these nodes in NATO's defense of the North Atlantic ranged from Bermuda in the west and Sigonella in Italy, far into the Mediterranean, and from Iceland and Norway in the north to the Azores in the south. Keflavik, however, stood out as a hub among the other spokes for North Atlantic ASW operations.[43] The bases and sensor networks ranging from Iceland and Norway in the north to the Azores in the south, along with the allied navies and air forces around the North Atlantic, were all required to work closely together to be able to successfully detect and track the increasingly quiet Soviet submarines and Soviet naval activity on the surface. Soviet SSBNs departing their bases on the Kola Peninsula is a case in point. U.S. and British SSNs normally tried to pick up the trail shortly after the SSBNs departed the Kola Peninsula, while MPAs from NATO member navies would maintain an overhead presence. As the SSBNs left the Barents Sea they would be tracked by Norwegian MPAs, which would later hand over the track to U.S. and British MPAs operating out of Keflavik and Kinloss in Scotland as the SSBNs approached the GIUK gap. South of the gap, Canadian MPAs from the Canadian east coast would pick up the track, to be joined later by flights from the Azores as the SSBNs steered in parallel to the U.S. east coast and toward the Caribbean. A typical continuous track of a Soviet SSBN in the North Atlantic could include aircraft from five NATO nations, operating from bases in six nations, ranging from northern Norway to southern Spain.[44]

The competition over the North Atlantic during the Cold War also gave rise to far-reaching technological cooperation between the NATO allies, including in the field of ASW. The American-produced P-2 Neptune MPA was exported to several NATO nations around the North Atlantic, including Canada, France, Britain, and the Netherlands. The Neptune's replacement, the P-3 Orion, similarly was adopted as the frontline MPA by the Netherlands, Spain, Portugal, Greece, and Norway. Often these procurements of American-made MPAs came with training of crews and maintenance personnel in the United States, which made for easier operational cooperation among the allies. In order to further extend SOSUS and other sensor networks into the Norwegian Sea and the Barents Sea, the United States cooperated with Norway in the 1960s to build additional listening posts and defensive barriers. As part of this cooperation Norway was also provided with P-3 Orions by the United States on favorable terms. Having Norway take responsibility for patrolling and tracking Soviet submarines and naval movements in the Barents Sea, or at least the efforts that were clearly visible to the Soviets, also had the additional benefit of not antagonizing Moscow more than absolutely necessary.[45] The United States also played an active role in helping buck up the ASW capabilities of individual allies during faltering moments. The Dutch decision in 1968 to get rid of the Netherlands' only ASW carrier (with a complement of both fixed-wing and ASW helicopters), which had previously served as an aircraft carrier, left a gap in the Netherlands' airborne ASW capacity. It was plugged with the procurement of P-3 Orions, and the U.S. Navy played a major role in advising the Dutch on how to set up their MPA organization, maintenance, and training procedures and design of infrastructure fit for the purpose of supporting MPA operations.[46]

Driven by the United States, NATO also established a NATO Undersea Research Center (NURC) to pool and share expertise across the alliance in regard to oceanography, underwater acoustics, and ASW technologies. Located in Italy, the NURC performed, among other things, a number of "military oceanographic campaigns" to better understand the subsurface environment in and around the GIUK gap, the Norwegian Sea, and elsewhere. The NURC research ships were often shadowed by Soviet intelligence-gathering ships.[47] There were also examples of exclusively European efforts of cooperation around ASW during the Cold War.

Along with the U.S.-produced P-3 Orion, the Atlantique Breguet served as the MPA workhorse of the mid–Cold War until it ended

in the early 1990s. In 1957, France introduced the concept of a pan-
European MPA that would be able to serve in ASW and other roles
in many nations' air forces and navies. Six NATO nations joined the
French-sponsored MPA study group, and the first Atlantique flew
only two years later, in 1959. Ultimately, only France, Germany, the
Netherlands, and Italy ended up procuring Atlantiques for their
ASW forces, with Canada and the United Kingdom opting out of the
consortium.[48] Nearly ninety Atlantiques were produced between the
years 1961 and 1987, and, as the Cold War receded, a handful were
later sold to the Pakistani navy. The Atlantiques served reliably during the Cold War, with relatively few losses of aircraft, although one
Dutch Atlantique was dramatically lost over the Atlantic off the west
coast of Scotland in 1981 as it was monitoring the Soviet aircraft
carrier *Kiev* in blizzard conditions.[49] While the French dream of a
common NATO MPA was not fully realized, the Atlantique must
be rated as a considerable success for pulling together four nations
to develop, procure, and operate such an advanced system. International defense equipment cooperation is a challenging enterprise
under the best of circumstances, and recent history is filled with
examples of projects that failed due to inadequate budgets, divergent requirements, national pride, and domestic politics. In 1980 the
French navy ordered a new and upgraded version of the Atlantique,
but this time without additional European partners.

In spite of the close military, political, and technological cooperation and the broadly shared goal of defending Europe from Soviet
aggression, it was not always easy to keep the NATO allies together
in the North Atlantic during the Cold War. On three occasions
between 1958 and 1976, Iceland and Britain jostled over fishing
rights in the North Atlantic. The so-called Cod Wars began when
Iceland extended its fishing zone to two hundred nautical miles
from its coast over fears that the fish stocks in the waters were being
exhausted by fishing fleets from the United Kingdom and Germany.
This naturally brought the little nation into conflict with Scottish
fisheries. Icelandic coast guard vessels cut the lines of British fishermen who had trespassed on what Iceland insisted was an area for the

exclusive use of Iceland fishing fleets. British and Icelandic trawlers also frequently rammed each other during the third Cod War. All in all, the Royal Navy deployed some twenty-two frigates and additional supply ships and tugboats to protect the British fishing boats as they sought to fish for cod in the extended exclusive zone claimed by Iceland.[50]

Very few injuries, and only one death, resulted from the third Cod War. The dispute between the two NATO members Iceland and Britain, however, still had strategic ramifications for the unity of NATO and threatened to seriously impede the United States' and NATO's presence and ASW operations in the North Atlantic. In a last-ditch attempt to keep its two-hundred-nautical-mile zone of exclusive fishing rights, the government of Iceland froze its diplomatic relationship with the United Kingdom and threatened to withdraw the country from NATO and, possibly, end access to the Keflavik airbase, although the Icelandic government never made the latter point explicitly. If access to the Keflavik base were lost, it would remove one of NATO's key nodes in the North Atlantic network and a crucial leg in the airborne ASW triangle that had been formed by Keflavik, Andoya in northern Norway, and Kinloss in Scotland. Only a NATO-facilitated round of mediations finally settled the matter with restored diplomatic ties between Reykjavik and London, and with Iceland still firmly within NATO.[51]

Iceland's bilateral relationship with the United States was also not without its turbulence during the Cold War. While Iceland had departed from its tradition of neutrality with the 1951 defense agreement, the ethos of seeking to remain outside great-power politics remained alive in Icelandic domestic politics, an understandable instinct that is not uncommon among small states in Europe and elsewhere.[52] In 1956 and 1974 there were serious moves to eject the Americans from Keflavik, which was only avoided after negotiations at the most senior levels of government.[53] Of course, Iceland was not the only NATO country with at times a tense relationship with the American superpower and the de facto leader of the alliance during the Cold War period. Indeed, Iceland's size and small population, along with the fact that the nation does not maintain a military force, meant that in many circumstances Iceland's position on a number of issues within NATO was of little consequence in comparison to those of the United Kingdom, France, Italy, and other major European NATO members. But its position as a toehold in the middle of the North Atlantic gave the small country real weight.

The occasional tension between Washington and Reykjavik was at times exploited by a Soviet Union eager to see NATO's control over the North Atlantic softened up. Public protests just outside the Keflavik airbase, which happened frequently throughout the Cold War and which often were driven by left-leaning Icelandic groups, were suspected of being encouraged and perhaps supported by the Soviet Union.[54] Moscow also looked to politically dislodge NATO from other parts of the North Atlantic during the Cold War. Moscow approached Norway, a founding NATO member, in 1956 to explore how the Svalbard Treaty could be modified to give the Soviet Union equal standing on the islands. The Svalbard Treaty gave Norway sovereignty over Svalbard, but it also gave the signatory nations (including the Soviet Union) commercial access to the islands and the surrounding waters, all in exchange for a commitment that Svalbard would remain without a permanent military presence. Still, Soviet concerns remained that Svalbard could provide a geographical toehold for military forces and create a new, and narrower, choke point that could be used to control access to the North Atlantic well north of the GIUK gap.[55]

Maritime Close Encounters and Incidents in the North Atlantic

The increased presence and activity of U.S., allied, and Soviet navies in the North Atlantic and other seas during the Cold War also led to close encounters between submarines, ships, and aircraft. Operations in close proximity to one another did not happen by accident or happenstance. Instead, they were done for the purpose of testing the political will or the behavior at sea of the other side, monitoring exercises, or for gathering intelligence and collecting signatures of surface ships, submarines, and aircraft.

While most encounters passed without incident, others led to losses of both lives and equipment and to tense moments at the political level, a trend that grew from the late 1960s on as the Soviet navy became more active and sailed farther afield. In late May 1968 the carrier USS *Essex* was conducting ASW training in the North Sea and was subject to a flyby at close range by a Soviet Tu-16 Badger bomber, which at one point flew so low that it was below the deckline of the American carrier. After passing the carrier the Badger pilot began to come around for another pass when it dipped its wing into the water and lost control of the aircraft. The *Essex*'s rescue helicopter, already airborne—as was and is routine during carrier launch and recovery operations—raced to the scene of the crash some two miles away from the *Essex* but found no survivors. At the same time, the *Essex* sent flash message traffic to Washington with information about the accident, as there was a real fear that the Soviet Union would mistake the crash as a U.S. shootdown of the Badger with potentially grave consequences.[56] Two years later the British carrier HMS *Ark Royal* was participating in an exercise when it collided with a Soviet destroyer that was monitoring the exercise. The *Ark Royal* was slightly damaged in the collision, while two Soviet sailors were lost overboard in the incident and were never recovered.

Other U.S. allies and NATO members in the North Atlantic were not immune to close encounters either. In 1971 a Norwegian P-3 Orion shadowing a formation of ships from the Northern Fleet in the Barents Sea was fired upon by a Russian warship, although it was clear that the Soviets were using training ammunition and not live rounds. In 1987, over international waters northeast of Murmansk, a Soviet Su-27 Flanker flew so close to another Norwegian P-3 that the Flanker's tail fin was sheared off by one of the P-3's propellers. The Norwegian P-3 immediately headed for home and was met on the way back by Norwegian F-16s, which were sent up to escort the MPA back to base.[57]

The major NATO maritime exercises of the Cold War in the North Atlantic also drew intense Soviet interest, and sometimes even short-notice reciprocal exercises by the Northern Fleet in the same maritime region. For example, NATO's September 1979 Ocean Safari exercise, which was held off the coast of Norway and specifically exercised the ability to protect and defend reinforcement shipping coming across the Atlantic, was met at sea by a Soviet task force from the Northern Fleet made up of cruisers and destroyers sent out on a no-notice exercise in response to Ocean Safari.[58]

Both the United States and the Soviet Union were concerned that such incidents at sea could lead to serious misunderstandings and unintended escalations. Thus in the early 1970s representatives from both the U.S. and Soviet navies met to hammer out an incidents-at-sea agreement that would to some degree regulate the behavior of each side during and after an incident at sea and how the two navies should operate when in close proximity to one another.[59]

Sometimes accidents or incidents involving only one actor also contributed to increased tensions. In May 1978 another Tu-16 Badger crashed on the small island of Hopen, off the Svalbard archipelago, during what was likely a reconnaissance mission. This incident happened during a period of increased Soviet air activity in northern Europe, with Denmark, among others, having to scramble fighter jets to intercept monthly incursions into national airspace. Tensions also rose further between Moscow and Oslo after the Badger crash, as the Norwegians resisted calls to immediately turn over the Badger's recovered black box. Press accounts from the time indicate that the Norwegian government intended to go over the data from the black box first before returning it to Soviet authorities.[60] In 1985 a cruise missile fired from a Soviet submarine during an exercise in the Barents Sea veered seriously off its set course and crossed into Norwegian airspace, overflew the northern tip of Norway, and ultimately landed in a Finnish lake.[61] A Finnish team located the impact area and gathered up the scattered missile body and returned it to the Soviet Union, presumably after carefully poring over it to glean technical intelligence about its performance and design.

One of the highest-profile incidents in the North Atlantic region during the Cold War was the sinking of the Soviet *Mike*-class submarine *Komsomolets* in April 1989. The *Komsomolets* was the only submarine in its class and had been launched by the shipyard Sevmash and delivered to the Northern Fleet in 1984. The *Komsomolets* had a titanium hull, which meant that it could operate much deeper than U.S. and other Western submarines. It also meant that the *Mike* class would be able to defeat some of the sensors deployed to find Soviet submarines, such as the magnetic anomaly detector, which had been installed on board P-3s. While the *Komsomolets* was intended to serve as a trial bed for the Soviet Union's next generation of SSNs, it did have full operational status with the Northern Fleet. While operating at around 1,100 feet on April 7, 1989, in the Norwegian Sea some 110 nautical miles southwest of Bear Island, a fire started in the engine room. The fire proved impossible for the crew to fight successfully, and the submarine reached the surface only eleven minutes after the fire had started. The crew began to abandon the submarine, and ultimately it sank a few hours after first reaching the surface. More than half the crew perished, either in the ultimate sinking of the submarine, while fighting the fire, or from hypothermia while awaiting rescue.

Given that the *Komsomolets* was nuclear-powered and had nuclear weapons on board when it sank, international concern was raised around the incident. At the time there was no way of knowing if the reactor or the weapons had been damaged during the incident. Under intense pressure from Norway, the Soviet government agreed to search for the submarine, and they located it some two months after the sinking, at a depth of nearly six thousand feet and some 125 nautical miles southwest of Bear Island. The Soviet Union insisted at the time that the risk of spills or nuclear hazards was negligible. The *Komsomolets* was not, however, the only Soviet nuclear submarine that was lost under controversial circumstances. In 1986 a *Yankee*-class submarine from the Northern Fleet sank east of Bermuda after a fire and flooding in the missile compartment, and there were allegations made at the time that the fire had started after a collision between the *Yankee* and a U.S. submarine, an allegation later disproved. These accidents, along with others ashore in the Northern Fleet area, not only caused concerns to rise within NATO but also signaled that while the rapid rise and growth of the Soviet navy was impressive, all was not well on the other side of the Iron Curtain. Many of the Soviet naval mishaps could be attributed to poor training and low technical proficiency among Soviet crews.

The End of the Cold War and a Changing U.S. Strategy

As the Cold War entered its last decade in the 1980s, the barrier strategy developed by NATO and the United States for the North Atlantic, leveraging the GIUK gap and using the aforementioned network of bases and sensors, began to change with the introduction of the U.S. forward maritime strategy under the Reagan administration. The strategy was based on forward operations north of the GIUK gap, with U.S. and allied submarines, carriers, and destroyers to strike at key targets on the Kola Peninsula. This strategy was informed by the growing recognition of the importance of the bastion system on the Kola Peninsula to Russia as a safe haven for Soviet SSBNs and the Soviet second-strike capability, as well as the need to ensure that NATO's northern flank would not be left indefensible at the initial stages of a war between NATO and the Soviet Union.[62] In addition, during war the forward maritime strategy was intended to put pressure on the Soviet Union's northern areas in order to force the Soviets to divert resources and attention from the envisioned central front on the plains of Europe. The new strategy was first implemented with the exercise Ocean Venture in 1981, which sent U.S. and allied naval forces, including 250 ships and submarines and 1,000 aircraft, well into the High North. Other exercises followed that intended to signal to the Soviet Union that Soviet naval power would be destroyed close to home and that land targets well inside Russia were now within reach of the U.S. Navy.[63]

The forward maritime strategy was enthusiastically advanced by the U.S. Navy and the Reagan administration, but it was not without its critics. Concerns were raised that this strategy would leave the crucial transatlantic lines of communications undefended and that the United States and its allies lacked the numbers and capabilities to sally into the Barents Sea and attack the heavily defended Kola Peninsula.[64] NATO allies in the North Atlantic region raised concerns that the forward maritime strategy was too offensive, risked needless provocations, and would contribute to rapid escalation during a crisis.[65]

The new maritime strategy was, however, never tested in real life. The third battle of the Atlantic never turned into a shooting war, as the European continent remained peaceful throughout the Cold War through deterrence, both conventional and nuclear. The new maritime strategy did, however, seem to shift the Soviet calculation about its chances of winning a potential war in Europe, thereby contributing to deterrence. The contest ended with the economic and political collapse of the Soviet system, rather than with a war that could have consumed not only the combatants but much of the world. Following the end of the Cold War, the North Atlantic region was poised for significant change, not only in terms of great-power relations and defense and security, but in economic and political terms as well.

CHAPTER 6

Lessons from the
Three Battles of the Atlantic

THERE ARE A NUMBER OF enduring lessons to be drawn from the intense naval competitions in the North Atlantic during the twentieth century, which remain relevant today and need to be considered as the North Atlantic once again turns into a competitive maritime space.

THE NORTH ATLANTIC OPERATING ENVIRONMENT IS VAST.
The North Atlantic ranges from the northeast and the Barents Sea in the Russian Arctic to the North American east coast, which accounts for nearly 10 percent of the world's surface. This puts an emphasis on forces and platforms that can operate across long distances, such as MPAs, access to basing in the broader region, or a burden-sharing arrangement where different allies focus on a specific area of the North Atlantic.

EVEN IN MARITIME CONFLICT THE LAND STILL COUNTS.
With such a vast operating environment, pieces of land are important for basing ASW forces and intelligence, surveillance, and reconnaissance (ISR) assets that can help detect, track, and attack naval forces,

especially submarines. Pieces of territory such as Greenland, Iceland, Scotland, northern Norway, and the Azores are especially important and will be of high value to both sides in a conflict over control of the North Atlantic.

CONTROLLING THE NORTH ATLANTIC IS A TEAM EFFORT.
Each successful attempt at controlling the North Atlantic and defeating naval challenges has been the result of a multinational effort at both the political and military levels. The side with no allies in and around the North Atlantic, whether it is Germany in the two world wars or the Soviet Union during the Cold War, has never been able to overcome the combined and coordinated resources and efforts of an alliance, be it the rapidly put together alliance during World War II or the more deliberate and elaborate NATO effort during the Cold War. And keeping the allies together is not always an easy thing. Witness the British and American frustrations with the Soviet Union's ability to contribute to effective ASW in the Barents Sea as German submarines and airplanes preyed upon Allied convoys during the effort to resupply the Soviet Union. Or see how the controversy over cod fishing between Britain and Iceland put access to the crucial Keflavik airbase at risk. The introduction of a new more offensive U.S. maritime strategy for the North Atlantic in the early 1980s also caused some unease among the NATO allies in northern Europe. The perspective of each ally is valid and reasonable from its own point of view, but these perspectives have nevertheless generated friction that has complicated and challenged the ability to keep the North Atlantic open during crucial times for American and European security.

This is not to say that go-it-alone approaches have worked in the air and ground domains of warfare in Europe during the twentieth century. There has been close cooperation with allies there too, but the North Atlantic domain has produced unique forms of cooperation over long periods of time that are different from those found ashore and in the air.

ASW IS FRUSTRATING, TEDIOUS, AND LONG-TERM WORK.
In all the battles of the Atlantic, defeating attempts to close the maritime domain with submarines has proven to be extended and frustrating work, often done through trial and error. It also deserves to be pointed out that each competition has been a close-run thing; no side has completely dominated the contest. Indeed, in the three battles of the Atlantic during the twentieth century, the opponent seeking to close the North Atlantic has innovated and been close to overtaking the defenders in technical terms, only to be ultimately defeated by the larger war effort. Also, ASW is resource-intensive and draws significant numbers of ships, aircraft, and submarines away for long periods of time from other missions and tasks elsewhere. Indeed, tying down valuable ships and aircraft could very much be part of a subsurface warfare strategy by an opponent employing submarines.

THE COMPETITION IS TECHNOLOGY-INTENSIVE.
The twentieth-century competitions in the North Atlantic were all technology-intensive efforts, on both sides of the conflict. The competitions there drove the development of sensors, propulsion systems, platforms, command-and-control systems, and weapons. Often these technologies were at the cutting edge of technological development at the time and relied on a number of technical and scientific fields, ranging from computing to oceanography. The enduring military dynamic of a competition between capabilities and their counters is also readily apparent in the contests for the North Atlantic, be it between sonars and quieting technologies, or radars and underwater endurance. This also means that any successful competition in the North Atlantic will be costly.

IT'S NOT ALWAYS ABOUT SINKING THE ENEMY.
It is easy to focus on the most dramatic aspects of the naval actions in the North Atlantic during the twentieth century; after all, the pursuit and sinking of a submarine provides for tense and compelling drama. ASW, however, is fundamentally about denying

the purpose of the submarine, whether it is to attack shipping or gather intelligence. Note the effectiveness, at least for a time, of the Dover Barrage during World War I or the ability of Allied forces to force German submarines during World War II to break off the attack and leave the area of operations.

BE CAREFUL WHAT YOU MEASURE.

Effective strategies for controlling, or denying control of, the North Atlantic depend on accurate assessment of both sides' intents, weaknesses, and strengths. This is not as easy as it seems. The late arrival of the effective convoy system during World War I had much to do with a widespread misreading of the statistics of how many ships were engaged in the resupply of Britain, which made the convoy concept seem impractical. The German submarine campaign during World War II may have been more effective if it had focused on attacking troop ships coming across the North Atlantic, which might have delayed and frustrated the American force buildup in the United Kingdom in preparation for the invasion of the European continent. Instead, the Germans used general tonnage sunk as the metric of success, giving German commanders little incentive to distinguish between the targets they pursued. The U.S. reassessment of Soviet naval strategy and capabilities in the late 1970s led to a wholesale re-evaluation of U.S. strategy in the North Atlantic toward the end of the Cold War and to a new emphasis on operating in the far North Atlantic and the Barents Sea, rather than maintaining a barrier defense around the GIUK gap.

PART II

Peace

THE END OF THE COLD WAR opened a new era in the North Atlantic, with both new challenges in the form of the rapidly decaying Soviet navy and new opportunities in the shape of emerging cooperation between the former Cold War enemies. It also brought attention to the growing economic importance of the North Atlantic, and in particular the region north of the GIUK gap. These developments also brought new actors to the Atlantic region. While the next nearly three decades would serve as an interlude of peace and cooperation in the North Atlantic, the period would also help shape the outlines of the emerging new contest between Russia and NATO in the North Atlantic that started in 2014.

CHAPTER 7

Rust and Nuclear Waste

The Collapse of the Soviet Union and the Northern Fleet

THE END OF THE COLD WAR and the breakup of the Soviet Union turned into a near total social and economic collapse for Russia. The early 1990s was a period of economic and political shock therapy, as the massive Soviet state was disassembled and formerly government-owned enterprises were privatized and exposed to global competition. As Russia felt its way toward a market-oriented economy and an open political system, unemployment spiked, public finances crumbled, and economic growth was nonexistent. By 1999 Russia's GDP had shrunk by 40 percent in comparison to 1989. In contrast, the U.S. great recession that started in 2008 only lasted for two years and saw U.S. economic output shrink by less than 6 percent. The Great Depression in the United States, from 1929 to the late 1930s, saw a drop of around 20 percent in America's economic output, a major decline indeed but still far from the contraction experienced in the former Soviet Union in the 1990s.

This also had an immediate and long-term impact on the Russian military, which had formed the great core of the fighting forces of the Soviet Union. Between 1988 and 1994, Russia's share of global military expenditures dramatically dropped from 21 percent to only a little over 4 percent. Lack of funding led to severe cutbacks in the size of the Russian military, going from 5.3 million soldiers, sailors, airmen, and officers in 1985 to a little more than 1.2 million in 1996, while maintenance, readiness, and, perhaps even more importantly, good order, discipline, and morale plummeted.[1] This poor state of affairs was widespread across the Russian military and included the Russian navy and the Northern Fleet, which had experienced such quick growth and played such a pivotal role during the Cold War. And the Russian navy took an especially large share of the cutbacks during the early post-Soviet period. During the 1990s the Russian navy's portion of the defense budget dwindled from 23 percent to only 9 percent. The navy's personnel strength declined to only around 200,000 and some 190 naval institutions and units were disestablished.[2] At the same time, the Northern Fleet's submarine force shrank from 180 submarines down to only 42.[3] Some of the laid-up submarines served, while waiting for decommissioning, as living quarters for sailors forced to live there due to the decrepit state of the living quarters ashore that had not been maintained for lack of funds.[4] Days at sea for the Northern Fleet's warships and submarines fell sharply, including for the SSBNs with the Northern Fleet. This not only eroded the Northern Fleet's overall readiness and operational nuclear posture, but further ate away at the skills of the Russian navy's crews, as they had few opportunities to practice their trade at sea.[5]

The low morale, erratic pay, and disrepair of the Russian navy also took its toll on individual sailors and officers, a state of affairs that on a few occasions took a tragic or near catastrophic turn. On September 11, 1998, a nineteen-year old Northern Fleet sailor named Aleksandr Kyzminikh, who was coming off guard duty, killed eight of his fellow sailors with an AK-47 service rifle he had stolen from another sailor on guard duty. Kyzminikh then barricaded himself in the torpedo room on board the *Akula*-class submarine on which he was serving. The submarine had live torpedoes on board, and Kyzminikh threatened to blow them up, along with the submarine. While the threat would have been difficult to actually carry out, it made for a dramatic image that gained some coverage in Western news at the time. The situation was only ended after Russian counterterrorism forces, who happened to be in the area for an exercise, were called in to storm the torpedo room. Kyzminikh died during the operation from either a self-inflicted gunshot wound or by rifle fire from the breaching team.[6] During the same period two groups of sailors and officers were arrested for stealing radioactive materials from their naval bases and submarine decommissioning facilities. Before their arrest they were looking for buyers for the materials on the black market. Around the time of the incident on board the *Akula*-class boat, the credible Norwegian-Russian nuclear watchdog organization Bellona wrote that "the social problems within the Northern Fleet have been of great concern throughout the last years. The soldiers have been serving without salary since May this year, and the current collapse in Russia's economy will not improve the situation for the soldiers in the north. Media reports have abounded in the last couple of weeks about lack of food at the Northern Fleet bases."[7] The decline of the Northern Fleet, and the collapse of the infrastructure built and maintained to support it, was also reflected in the broader society of the Kola Peninsula. The Murmansk and Arkhangelsk regions, where the bulk of Northern Fleet and its associated yards and facilities could be found, lost some 600,000 inhabitants between 1989, when the region had been a boom town fueled by military spending, and 2011, a decline of roughly a quarter of the population.[8]

By the early 2000s it was doubtful whether the Kola Peninsula, the great bastion of the Soviet navy and home of the Northern Fleet, the erstwhile Soviet Union's most capable fleet, really was of strategic consequence to the United States and the West more broadly anymore. Deterrent patrols with the SSBNs of the Northern Fleet were few and carried out with aging submarines of the *Typhoon* class. The air defenses on the peninsula had also crumbled, which meant that the Northern Fleet, in theory, was more vulnerable than ever to strikes against Russia's sea-based nuclear deterrent, even with conventional weapons such as the U.S. Tomahawk cruise missile. Many suggested that given the general decline of the Russian military, but in particular the inability of the now sclerotic Northern Fleet to threaten the North Atlantic sea-lanes and the vulnerability of the Russian SSBNs there, perhaps Russia was destined over the long term to abandon its northwest as a strategic area.[9] This would of course also mean that Russia would no longer be able to operate in the North Atlantic in any meaningful way, since the fall of the Soviet Union also meant that Russia had to retreat from many of the naval bases it had established in the frontiers of its empire. With the Baltic states regaining their independence, the Russian navy was forced to depart from its naval basing in Estonia, Latvia, and Lithuania, leaving Russia with only a maritime toehold in the Baltic in Saint Petersburg and in the Russian Kaliningrad enclave in the southeastern corner of the Baltic Sea, crammed in between the increasingly Western-oriented (and later full NATO members) Poland and Lithuania. The declining quality and quantity of Russia's Baltic Sea Fleet at those remaining bases also meant that the fleet lacked the power to push through the Baltic and into the North Sea and the Atlantic during wartime. Russia also had to give up its basing on the Crimean peninsula and accept a more precarious arrangement where the Ukrainian state leased the facilities back to the Russian Black Sea Fleet.[10] The Russian navy also lost important personnel training centers in both Estonia and Ukraine, which would prove costly and time-consuming to start anew inside Russia.[11] With the retreat from forward naval basing in former Soviet republics, Russia also lost many shipyards that were located in now independent nations, and with that some of the know-how about maintaining and building naval vessels slipped away too. This was sometimes an acute loss, as, for example, turbine engines for navy ships were supplied by factories in Ukraine. The dispersal of the naval industrial base across the Soviet Union had had a clear and practical logic during the Cold War: It made the naval industry more resilient, as the sudden loss of one production center due to attack could be made up by shifting production to another location. It also tied outlying parts of the Soviet empire more closely to its Russian center.[12] But with the breakup of the Soviet Union that logic backfired in a profound way on the Russian navy and Russia's maritime power more generally. Nevertheless, the intellectual know-how to design sophisticated submarines and warships remained with the prestigious Russian design bureaus, such as the Rubin Central Design Bureau for Marine Engineering in Saint Petersburg. The remaining naval industrial base inside Russia also fell on hard times during the post–Cold War period. The yards of Sevmash, the Russian naval contractor in Severodvinsk near Arkhangelsk on the White Sea, which had grown into the world's largest producer of nuclear submarines during the Cold War and had so impressed Admiral Rickover, was forced to diversify away from its sole focus on naval construction and nuclear vessels. By the mid-1990s Sevmash instead constructed cruise ships and container ships, along with more exotic platforms such as floating nuclear power plants. But this attempt at diversification was not easy. Over the years, several contracts were cancelled by disappointed international clients due to delivery delays, poor quality, and rising costs.[13]

A New Nuclear Threat

But the decaying Northern Fleet led to another form of security risk for the nations around the North Atlantic, and in particular those north of the GIUK gap. Poorly managed radioactive waste and loose nuclear materials from the rotting submarines of the Northern Fleet now threatened the environment of northern Europe and could potentially become a proliferation source for nuclear terrorism or rogue state actors.

During the decades of the Cold War, Russia's naval industry pumped out 245 nuclear-powered submarines and a total of 445 nuclear reactors for naval use. The large number of reactors highlights the Soviet and Russian preference to power its nuclear submarines with two reactors rather than one, which sets it apart from the nuclear submarine forces of other nations.[14] This left post–Cold War Russia with the monumental task of decommissioning them in a safe manner. A full decommissioning of a nuclear submarine is a massive undertaking that generates huge amounts of waste materials, including nuclear and other toxic substances that must be taken care of and processed. Along with nuclear materials, breaking up nuclear submarines can yield as much as one hundred tons of lead and around sixty tons of copper.[15] By 1995 some seventy-five nuclear submarines were awaiting decommissioning with the Northern Fleet on and around the Kola Peninsula, in close proximity to northern Europe. And even laid-up nuclear submarines require attention and resources in order not to dangerously deteriorate before they are safely broken up and processed. A skeleton crew of roughly 40 percent of full manning is required to keep a nuclear boat safe and secure at pier, even if it is never intended to go to sea again.[16] The massive backlog of nuclear submarines awaiting full decommissioning while at pier and without ongoing maintenance in the Northern Fleet caused Western experts to fear that many of them would sink before their nuclear fuel and reactors were removed.

The 1990s Russian navy and the shipyards, starved for funds and qualified personnel, faced real difficulties in taking on this technically difficult, tedious, and time-consuming work. Russia lacked the service ships needed to defuel the submarines, the infrastructure needed to safely transport the fuel away for storage was crumbling, and the long-term storage sites for the cut out reactors had not yet been built. Sevmash and the other shipyards in Severodvinsk along with four others on the Kola Peninsula were thrown at the problem of scrapping the Northern Fleet's old nuclear submarines, but the pace turned out to be slow. By 1995 only six submarines had been fully decommissioned; at that pace it would take several decades for the yards to clear the Northern Fleet's log of more than seventy nuclear submarines.

The challenge of disposing of Russia's Cold War nuclear submarine fleet was not only due to technical issues. Administration, or lack thereof, played a major role too. The shipyards in Russia's north complained repeatedly that Moscow failed to pay for much of the decommissioning work that had already been performed. In 1994, less than 25 percent of the funding allocated for decommissioning work was actually handed over to the yards and agencies performing the work. And decommissioning a nuclear submarine is expensive. A 1993 estimate pegged the cost at roughly $150 million per submarine in today's value, meaning that the overall cost of processing all of the Northern Fleet's laid-up nuclear submarines alone would approach $12 billion in today's value, a daunting figure for a Moscow that had already seen the Russian economy implode and had reduced defense spending by nearly 70 percent between 1992 and 1999.

The dangers created by the slow pace of decommissioning and the lack of funds were not theoretical. In October 1995 the meltdown of a Russian submarine reactor near Murmansk was just barely avoided. The submarine in question was awaiting decommissioning at pier and still had its fuel and reactor on board. The reactor and fuel were cooled using electricity from an onshore source, which suddenly was cut off by the power company because the navy had not paid its electric bills for many months. The power used to cool the submarine was only restored after the Northern Fleet dispatched an armed team to the power station to compel the technicians there to turn the power back on.[17]

The Russian struggles to decommission the Cold War submarine fleet got high-profile attention from the West, which was concerned with the environmental implications as well as the risk of nuclear materials making their way on to the global black market. Both Norway and the United States provided assistance to help build onshore long-term storage facilities for the nuclear waste on the Kola Peninsula.[18] In 2012 Russia began the decommissioning of the Northern Fleet's last Cold War–era nuclear submarine.[19] This signified the beginning of the end of a colossal and complex effort that had begun more than two decades earlier and that had, in effect, become an international effort that brought together the erstwhile opponents during the Cold War to help prevent an environmental security crisis in the Arctic and the North Atlantic. Still, crucial work remained, including solving the problem of long-term storage for the submarine reactors and the spent fuel, along with safely scrapping the many service ships that unloaded the nuclear fuel from the submarines, now themselves contaminated.[20]

The Nadir of the Northern Fleet

Russia's enormous problems with processing the toxic and dangerous legacy of its Cold War–era submarine fleet garnered real attention in both the United States and Europe, and U.S.-Russian cooperation in this field was viewed as a real opportunity for confidence building and establishing habits of cooperation between the two erstwhile enemies. It was not, however, without its critics. The U.S. Navy voiced concerns about bringing Russian engineers to U.S. submarine yards for technical training, and they also pointed out that Russia was still expending funds on building new classes of nuclear submarines instead of dedicating all of its own resources for decommissioning work.[21] And, in fits and starts, the Russian government did indeed try to reawaken its decrepit navy in the late 1990s, with much of the effort focused on the Northern Fleet. Work continued on new *Oscar-II* SSGNs, along with the development of the *Severodvinsk* class of nuclear multipurpose attack submarines. But the early attempts to reconstitute the Russian navy also led to one of the largest modern catastrophes of the Russian navy and the Northern Fleet: the loss of the *Oscar-II* submarine the *Kursk*, a calamity that would leave many thinking that the Russian navy would never truly rise again.

The *Oscar-II* class were some of the largest cruise missile submarines ever built, stretching nearly 500 feet in length and 60 feet in width. Construction on the *Kursk* began in 1990, one of the last submarine construction projects begun in the waning days of the Soviet Union. Designed to attack U.S. aircraft carriers, the *Kursk* was delivered to the Northern Fleet as an active SSGN in the first days of 1995 and formed part of Putin's early attempt to resurrect the Russian navy after a decade of decay. The *Kursk* was dispatched from the Northern Fleet through the GIUK gap and into the Mediterranean during NATO's Operation Allied Force in 1999, the Western effort to bomb Serbia into submission over the ethnic cleaning of Albanians in the then-Serbian province of Kosovo. While in the Mediterranean the *Kursk* observed NATO naval operations in support of Allied Force and was in turn pursued by NATO ASW assets, including P-3 Orions. The deployment was seen in Moscow as a first sign of the

Russian navy's return to the high seas, and the deployment represented the first time in more than a decade that a Russian submarine had operated in the Mediterranean. In the year 2000 the Northern Fleet planned to follow the *Kursk* deployment to the Mediterranean with one by Russia's sole aircraft carrier, the *Kuznetsov*, following a decade of frustrating sea trials and delays for the ship. But Russia's turn-of-the-century naval resurgence came to a screeching halt in the summer of 2000, with the dramatic accident on board the *Kursk*, which in many ways had led the way for the Russian navy's growing operational ambitions.

On August 10 the *Kursk* departed its pier at Vidyayevo on the Kola Peninsula and headed out into the Barents Sea to join an ongoing exercise with more than thirty surface ships and additional aircraft. That the *Kursk* was part of this major naval exercise was not secret. A Norwegian P-3 had photographed the *Kursk* while on the surface in the Barents Sea in mid-July, as she was preparing to take part in the exercise.[22] On August 11, the *Kursk* was to fire one of its SS-N-19 Shipwreck antiship cruise missiles, and the next day was to engage the cruiser *Peter the Great*, playing an enemy aircraft carrier during the exercise, in a torpedo attack using a dummy warhead. And while this was an exercise during peacetime, the *Kursk* still carried weapons with live warheads on board. The exercise that the *Kursk* was participating in was watched closely by Western navies. The submarines USS *Toledo* and USS *Memphis* were in the area (this was the *Memphis*'s second deployment to the Barents Sea in a year), along with the Royal Navy's HMS *Splendid*. About two hundred miles away was the USNS *Loyal*, collecting acoustic intelligence, while the Norwegian government had dispatched its signals intelligence ship the *Marjata* to the area.

The launch of the *Kursk*'s Shipwreck missile went without incident on August 11, and the next morning the crew of the *Kursk* began to prepare for the combat exercise against the *Peter the Great*. Shortly before the exercise start time of 11:30 in the morning the crew began to load the torpedo. During the handling of the torpedo, it exploded in the forwardmost compartment of the submarine, with the explosion ripping through the submarine, causing catastrophic damage to the

hull and instantly killing many of the crew. A second, larger explosion
followed quickly thereafter when some of the other live weapons on
board the submarine detonated. The two explosions were detected
by the trailing U.S., British, and Norwegian units, including by the
USS *Memphis*. The *Memphis* was a *Los Angeles*–class boat that had
been modified and upgraded in 1989 and had the ability to operate
unmanned underwater vehicles and new towed sonar arrays to extend
her intelligence-gathering capabilities, and she was some sixty-five
nautical miles away from the *Kursk* at the time. As the *Kursk* sank
from periscope depth in roughly three hundred feet of water, a small
band of survivors made their way into the aftmost compartment of
the submarine, which was largely undamaged. The surviving group
of sailors stayed alive for some time, but were unable to escape the
submarine due to a lack of training and equipment and an escape
hatch that had been left inoperable after leaving port. The group
ultimately perished after they tried to replace an air scrubber, which
came into contact with water or oil and caused a chemical reaction
that triggered a violent fire in the sealed-off compartment.

Not only was the loss of the *Kursk* a disaster for the Northern
Fleet and Russia's national leadership, the immediate response to
the accident would cause international embarrassment. Russia's own
submarine rescue system was woefully outdated and did not have the
means to attempt a rescue of the crewmembers of the distressed sub.
Indeed, the Northern Fleet's submarine rescue ship required calm seas
to safely hoist its submersible into the water, a condition rarely found
in the Barents Sea. The Russian government also refused offers of
help from the United States, the United Kingdom, Norway, Germany,
and Sweden, and it provided contradictory reports of the state of the
Kursk and the chance of finding any survivors. Even after accepting
help from the United Kingdom and Norway, the Northern Fleet was
reluctant to allow the Norwegian ship and the British submarine rescue
submersible to approach the *Kursk* and attempt to mate with it.[23]

In the aftermath of the *Kursk* accident, Russia's Northern Fleet
also made claims that the *Kursk* had collided with a U.S. or European submarine that was observing the exercise that the *Kursk* took

part in. Russian MPAs also sortied out over the Norwegian Sea in the days following the *Kursk* accident (where they were met by Norwegian F-16s), with the Northern Fleet alleging that they were tracking a damaged U.S. submarine limping out of the Barents Sea and toward the Norwegian navy's main base at Bergen. The *Memphis* did indeed arrive at Bergen shortly after the *Kursk* accident, but it showed no signs of damage and most likely called at Bergen to offload intelligence materials and personnel who would provide their own report to the U.S. Navy on what they had picked up about the *Kursk* accident.

The *Kursk* incident also proved a personal embarrassment for President Vladimir Putin, who seemed reluctant to return from his summer vacation by the Black Sea to deal with the growing emergency and calls for his personal involvement. It remains unclear if he failed to understand the gravity of the situation or if his military leaders simply did not convey to him how dire the situation really was. Putin finally traveled to the submarine base to meet with the family members of the *Kursk* crew ten days after the accident had occurred. The president of Russia was met by an angry crowd that verbally attacked him and the Northern Fleet admirals flanking him during the meeting. Parts of the chaotic meeting were broadcast on Russian TV, and the episode proved to be Putin's first public encounter with a crowd who questioned his leadership and ability to manage the Russian military. It would leave a permanent impression on Putin, who never again exposed himself to such nonorchestrated encounters with the Russian public.[24]

The *Kuznetsov* deployment to the Mediterranean was canceled in the aftermath of the sinking of the *Kursk*. Putin also removed the commander of the Northern Fleet as well as the commander of the Russian submarine force. Russia's minister of defense at the time was also shown the door. The Russian navy also worked to address its failing submarine rescue system, and Russian submarines would from time to time join NATO submarine rescue exercises in the years after the *Kursk* disaster.[25] During these exercises European submarine rescue vehicles would successfully mate with the participating Russian submarine and simulate a rescue of the crew, exactly in the fashion that had been offered by several nations to Moscow in response to the *Kursk* accident. The *Kursk* tragedy lingered for years as an image of what the great Soviet navy had become under the new system in a Russia cut down to size. It left many thinking that Russia's larger maritime ambitions had sunk along with the *Kursk* in the early 2000s.

CHAPTER 8

High North, Low Tension

As the Russian navy crumbled on the Kola Peninsula and else-where, and NATO took in new members and a permanent peace had seemed to set in across Europe, the North Atlantic region saw a new era emerge that would be characterized by increased economic activity, international cooperation, and the appearance of new play-ers from outside the North Atlantic region.

Regional Cooperation

During the 1990s and the first part of the twenty-first century, the maritime domain in the far North Atlantic and the Barents Sea became the arena for growing cooperation between the erstwhile Cold War opponents. Norway and Russia created the annual POMOR exer-cise, named after the term for Russia's historical Arctic settlers, which brought together Norwegian air and naval units with still operating warships and jets from Russia's Northern Fleet. The focus of the exer-cises rhymed well with the envisioned new world situation of failing states and nonstate threats, as well as West-East confidence building, and included boarding operations and search and rescue at sea. The POMOR exercises also included a strong symbolic element, with the exercise event usually starting around the annual Russian celebrations of the victory over Nazi Germany. At the time the POMOR exercises were viewed as the harbinger of more things to come in terms of naval cooperation between Russia and Norway in the Barents Sea. Commander Lars Saunes, the Norwegian lead for the POMOR exercise at the time, was quoted in 2011 as saying, "the last two-three years have been incredible when it comes to military cooperation with Russia. We trust each other." The commander of Russia's Northern Fleet was of a similar opinion when he commented on POMOR to the assembled media that "we are looking forward to closer cooperation with the Norwegian armed forces . . . we are making new steps all the time." As late as 2013 there were plans to further grow Russian-Norwegian military cooperation and expand the relationship to include information exchange and air and land exercises.[1]

The Norwegian willingness to find ways to cooperate with the Russian military should not, however, be confused with a national naiveté about Russia and its potential future trajectory or with a Norwegian ambition to embark on a course that would see it depart from the West and draw closer to Moscow. Small states, no matter how firmly ensconced in defensive alliances, must find a way to coexist on a daily basis with far larger neighbors, however unpredictable and potentially hostile they at times may be. This is especially true for regions such as the far North Atlantic and the High North, the European Arctic, where the harsh and austere conditions put a premium on cooperation to preserve lives and keep societies going. Besides, military exercises are a great way to get a close look at the ships, equipment, and personnel of the other participant. Exercises like POMOR therefore allowed the Norwegian navy a unique opportunity to closely observe the strengths and weakness of the Russian Northern Fleet under new management.

In that vein, the naval cooperation between Norway and Russia was also mirrored on the civilian side with Exercise Barents. Started in 1991, the exercise drew together coast guard ships and helicopters to rehearse responses to potential incidents in the opening and increasingly active Arctic, including search and rescue and oil spill prevention. Exercise Barents accelerated in the mid-2000s, changing from a biennial event to an annual exercise. The need for cooperation between the Russian and Norwegian coast guards was intimately related to the fall of the Soviet Union. With the weakened Russian state after the collapse of the Soviet empire, there was little in the way of monitoring the Russian fishing fleet that plied their trade out of the Kola Peninsula or of effectively enforcing their fishing quotas. It was thus in Norway's interest to find ways to regulate and manage the fishing of cod in the Barents Sea, a resource that was just as important to Norway as to Russia.[2]

NATO also began to cooperate with the Russian military in Murmansk in an effort to improve regional coordination to counter hijacked airliners, a real fear in the post-9/11 world. The air picture from radars in Murmansk were shared with those at the military command center in Bodø, Norway, which were then further shared with all NATO allies to get a full picture of the air domain in the far reaches of northern Europe. Ships and aircraft from the Northern Fleet and European nations also began to acquaint themselves with each other during international air or naval meets. In 1996, for example, a Russian IL-38 MAY MPA from the Northern Fleet flew into the United Kingdom to participate in the International Air Tattoo there.[3]

The Soviet *Mike*-class submarine *Komsomolets*, which had so dramatically sunk in the Norwegian Sea south of Bear Island in 1989 and caused concern across the North Atlantic region, also became a way to help establish a constructive relationship between NATO and Russia. In the early 1990s NATO established a scientific committee, made up of experts from both NATO and former Warsaw Pact countries, to study the effects and potential hazards of nuclear waste associated with the *Komsomolets* in particular, and the problem of military nuclear waste more broadly. The resulting report took great pains to be balanced and not accusatory so it could provide substance for continued military-scientific cooperation between NATO, Russia, and the other former Warsaw Pact nations.[4]

With Russia's Northern Fleet crumbling and a tentative partnership emerging between Russia and NATO, more hard-nosed and large-scale exercising by NATO naval forces in the North Atlantic dropped off quickly after the end of the Cold War. In mid-1992 NATO held its last iteration of an annual amphibious exercise in the Norwegian Sea, and it was a fifth smaller than originally envisioned. At that time NATO announced that the annual exercise was going to take a pause and return to the region in 1995.[5] It never did. Instead, NATO's limited presence in the North Atlantic focused on softer maritime security issues through the exercise series Dynamic Mercy, an annual search-and-rescue-at-sea training event that alternated between being held in the Baltic Sea and the North Atlantic. Instead of ASW and amphibious landings, the Dynamic Mercy exercise brought together regional rescue coordination centers and civilian emergency response units for drills focused on, as was the case in 2012, a plane crash in the far North Atlantic and an earthquake response on Jan Mayen island.[6]

The Eagle and the Bear: Wary Friends in the North Atlantic

The United States also made its first steps to deepen military cooperation with Russia in the far north. On July 1, 1992, the USS *O'Bannon* and the *Ticonderoga*-class cruiser USS *Yorktown*, which had been part of a dramatic at-sea encounter with a Soviet frigate in the Black Sea in 1988, sailed into the Russian Northern Fleet's base at the highly restricted town of Severomorsk to begin the first joint U.S.-Russian naval exercise, dubbed Northern Handshake. The exercise was rudimentary, but it provided an opportunity for the U.S. and Russian sailors to get to know each other and start building trust between the two sides. The U.S. sailors were allowed to also take in the town, an unthinkable thing just a few years earlier, and they exchanged pleasantries with the Russian residents of Severomorsk.[7] U.S.-Russian naval engagement accelerated after that, and the exercises turned to more advanced tasks. The U.S.-Russian naval exercise series Northern Eagle began in 2004, when the *Ticonderoga*-class cruiser USS *Hue City* and the Russian warships *Admiral Levchenko* and *Severomorsk* practiced basic naval interactions and mainstay maritime security tasks, such as maritime interdiction and choke point escort in the North Sea. During the exercise the crew of the *Hue City* marveled at, among other things, the skills and discipline of the Russian firefighting parties that were used during a damage-control drill, but they also noted that Russian surface warships came with little in the way of automated firefighting systems, such as halon dispensers, that would help the sailors on board to control a fire.[8] The Northern Eagle exercise was later expanded to include the Norwegian navy, and it was moved north into the Norwegian Sea and the Barents Sea. The 2012 Northern Eagle exercise featured the return of a U.S. warship to Severomorsk for a port visit, on this occasion conducted by the *Arleigh Burke*-class destroyer USS *Farragut*.[9] During the same period Russian warships visited U.S. naval bases too, including Mayport and Norfolk.[10]

Even the sinking of the *Kursk* in 2000 helped generate new cooperation between the United States, Russia, and other partners, in this instance around submarine rescue coordination. This cooperation was later formalized in the International Submarine Escape and Rescue Office, which is still active as of this writing, first headquartered in Norfolk, Virginia, and later moved to the United Kingdom.[11]

The North Atlantic and Arctic-focused cooperation between Russia and the United States and its allies grew ashore and at the political-military level as well. United States European Command, under the leadership of Adm. James Stavridis, created the Arctic Forces Roundtable, which drew together senior military leaders from the Arctic nations, including Russia, along with the United Kingdom, France, Germany, and the Netherlands. The early iterations of the roundtable effort focused on soft security issues in the Arctic, such as climate change and search and rescue. But while naval and maritime cooperation between the erstwhile opponents grew apace in the 1990s and early 2000s, it was not without its hiccups. Suspicions lingered, and the tensions between NATO and Russia elsewhere would find expressions in the far North Atlantic as well. For example, NATO's air campaign against Serbia in 1999, which was deeply unpopular with Russia, caused Moscow to cancel its participation in a planned NATO exercise in the Barents Sea.[12]

As the European security environment truly began to change with Russia's 2014 annexation of Crimea, there were early stirrings in the North Atlantic that the dynamic was about to change there too, even though it was far removed from the center of the events in Europe in 2014. In response to the Russian aggression, the Norwegian-Russian annual POMOR exercise scheduled for late 2014 was halted and has not resumed as of this writing. The exercise will likely not return until Russia changes its assertive behavior against NATO and gives up its claim to Crimea. Russia was also disinvited from participating in U.S. European Command's Arctic Forces Roundtable.[13] Bilateral coast guard exercises, however, continued between Norway and Russia in the Barents Sea. This interaction between coast guards was viewed by both sides as outside the bounds of the re-emerging geopolitical competition in Europe and direly needed to keep fishermen and oil workers safe in the forbidding maritime domain.

Economic Growth and Interests from Afar

As the North Atlantic turned from a zone of tensions and potential battle to one of peace and potential partnership between the two Cold War opponents in the 1990s, the Atlantic maritime domain also caught the attention of the energy industry in Norway, Russia, the United Kingdom, and other nations. The 1990s saw the eastern part of the North Atlantic and the adjacent seas become generators of real wealth for the surrounding nations in the post–Cold War era, in large part driven by offshore production of oil and gas. Modest oil and gas exploration in the North Sea and around the southern tip of Norway began in the 1960s, but then expanded northward into the Norwegian Sea and the Barents Sea. Oil and gas production in the North Sea and the Norwegian Sea garnered additional interest following the 1973 Yom Kippur War between Israel and a coalition of Arab states, which impacted global energy shipments. This was followed by an oil crisis after the Arab states involved placed oil export embargoes against Western states perceived to have supported Israel. This sent global energy prices spiking and spurred the development of alternative energy sources away from the Middle East. But while oil and gas exploration in the eastern North Atlantic began during the Cold War, platforms far out to sea came into full rate production only in the 1990s and early 2000s, a development in no small part driven by the increasing energy demand in the newly liberated economies of eastern Europe and the growing economies of Asia.

The growing globalization of the world economy, along with the rise of Asian powers, also contributed to an increasing internationalization of the far North Atlantic after the end of the Cold War. In 2013, China, Singapore, and India all gained observer status in the Arctic Council, an important intergovernmental body comprising the world's Arctic states (the United States, Canada, Norway, Denmark, Iceland, Sweden, Finland, and Russia) that deals with a range of Arctic issues, including development, environmental protection, and Arctic shipping. Much of the North Atlantic region, including northwestern Russia, northern Norway, Iceland, and the surrounding seas fall under the purview of the Arctic Council. And the growing interest in the North Atlantic region in China and elsewhere was unmistakable. Since the early 2000s China and other nations have engaged with the region through, among other things, scientific expeditions to Svalbard, investments in Iceland, and an interest in developing mining operations in Greenland.

As the waters in the North Atlantic and the adjacent seas have become an increasingly active space, it has led to mild forms of political and economic pushing and shoving between the nations and interests in the region. These tensions have sometimes emerged even between otherwise friendly nations. Snow crabs mysteriously began appearing in the Barents Sea in the mid-1990s, and the population has since experienced explosive growth.[14] This has proven an economic boon to both Norwegian and Russian fishermen in the region, but it has also drawn commercial fishing boats from other European nations. In early 2017 the Norwegian coast guard boarded the Latvian fishing vessel *Senator* and brought it into port, alleging that it had engaged in illegal fishing over the continental shelf by Svalbard, an area Norway claims jurisdiction over.[15] This was no isolated incident. It was a strong Norwegian signal in an ongoing dispute between Oslo and the

European Union (EU). This is not to say that higher level of economic activity will lead to military conflict. There are plenty of mechanisms for peaceful dispute resolution. Still, the tensions between divergent national interests in the pursuit of economic gain in the broader North Atlantic are unmistakable. Also, the increased economic activity in the North Atlantic region, especially north of the GIUK gap, brings a clutter of activity ranging from helicopters and ships servicing the oil platforms, scientific activities, pipeline maintenance, and fisheries; some two thousand vessels are licensed to conduct fishing in the Norwegian and Barents Seas. This in itself brings a need to monitor the region and ensure that undue interference does not occur.[16]

CHAPTER 9

Pirates, Terrorists, and the Hindu Kush

NATO at Sea in the Post–Cold War Era

THE END OF THE COLD WAR also began a profound transformation of NATO and the American engagement in European security. It would see the alliance drawn away from its North Atlantic focus, maritime roots, and deterrence as its reason for existing. And the United States encouraged this reorientation along the way. Operation Sharp Guard is instructive as an early example of the new roles taken on by the transatlantic alliance in the post–Cold War era.

By 1993 the civil war between Croats, Serbs, and Muslims in Bosnia had raged for nearly two years. Shocking footage from the fighting filled the TV screens of Europeans and Americans, and streams of refugees had begun to pour into western Europe at a rate not seen since World War II. The United Nations (UN) and the West tried to stem the violence by imposing an arms embargo on the warring parties, an embargo that was left to NATO to uphold and enforce. With a UN mandate in its pocket, NATO launched Operation Sharp Guard in the summer of 1993, with warships, oilers, and MPAs from thirteen NATO nations, including the United States, Turkey, the United Kingdom, Germany, and Canada. On the last Sunday of

April 1995, the Maltese-registered cargo ship *Lido II*, carrying fuel oil, was tracked by the NATO warships on station in the Adriatic off the coast of Montenegro (then a part of Serbia) when it suddenly changed course and headed toward the coast while claiming that the ship had an urgent leak in the engine room that necessitated making landfall on the embargoed coast. The on-scene British NATO commander cleared a U.S. destroyer and a Dutch frigate shadowing the *Lido II* to use disabling fire to stop the tanker. This authorization to use force was not used by the two warships' commanding officers, who were nervous about the ramifications of firing upon a civilian tanker in what was, as far as NATO was concerned, essentially a peacetime mission. Meanwhile, three corvettes from the Yugoslav navy darted out from the coast to meet the tanker and the hovering NATO warships, and they were only persuaded to break off the advance after a close encounter with the British frigate HMS *Chatham* and close overflight of Tornado fighter-bomber aircraft that had scrambled from their Italian airbase. *Lido II* was eventually boarded by Dutch marines, who made their way to the ship on board an Italian navy helicopter. The leak in the engine room was found to have been an act of sabotage and the *Lido II* was diverted to an Italian port for further inspection.[1]

Following the incident with the *Lido II*, Operation Sharp Guard continued in the same vein for another year, with NATO naval units boarding some six thousand ships in the Adriatic and intercepting a dozen arms smuggling attempts to former Yugoslavia over the three-year period of Sharp Guard. The operation proceeded with few major events, and the near shooting incident involving the *Lido II* is hardly remembered today. Both Operation Sharp Guard and the *Lido II* incident, however, were the first of things to come for the maritime forces of NATO in the following decades: the warships and submarines of the alliance would face pirates, terrorists, and smugglers under murky conditions given the label of "operations other than war," rather than the fleets of the now defunct Soviet Union. Most of these operations would also take place far from the North Atlantic and the other maritime spaces that have been the traditional operational spaces for NATO's member navies. With Operation Sharp Guard, NATO set a new course for itself at sea for the next two decades.

After the end of the Cold War, NATO changed profoundly, and left its posture as a defensive alliance of sixteen members fully focused on deterring against an attack from the Soviet Union and "keeping the Russians out, the Germans down, and the Americans in," as Lord Ismay, NATO's first secretary general, put it. Instead, NATO began a transition toward a more proactive role on the global stage, a process in no small part nudged forward by a Washington seeking a partner, and extra troops, for keeping the so-called Pax Americana across the globe. Twenty years after the conclusion of the Cold War, NATO counted twenty-nine members and was trimmed and primed for expeditionary operations in far-flung places. What happened to the alliance, America's allies in Europe, and the U.S. presence in and around Europe in between those two mileposts has had a profound effect on NATO's maritime posture and the capabilities that can be brought to bear in the now-emerging maritime competition in and around the North Atlantic.

Out of Area or Out of Business
The fall of the Soviet Union led to a breakdown in the established European security order, which in many places was managed peacefully but sometimes led to violence. The civil war in the Balkans was the first outbreak of major and sustained violence in Europe since World War II, and the UN task force, the United Nations Protection Force (frequently called UNPROFOR) deployed in Bosnia was quickly found wanting due to its restricted mandate and the mixed quality of the forces deployed as part of the task force.[2] NATO provided air strikes and embargo operations, such as the already mentioned Operation Sharp Guard, to apply pressure on the combatants to reach a cease-fire agreement, and once the conflict was settled, a NATO-led force (the Implementation Force) was inserted to help keep the peace and assist in rebuilding Bosnia. This became NATO's first major out-of-area operation, but it was

certainly not the last. In 1999 NATO was once again in combat in the Balkans in a bid to stop Serbian president Slobodan Milosevic's ethnic cleansing of the Albanian population in the tiny Serbian province of Kosovo (which later declared independence and is now a sovereign nation). After a ninety-day air campaign, which included NATO strikes from the sea from submarines, destroyers, and aircraft carriers, Milosevic finally buckled, and NATO once again put together a peacekeeping force to enter Kosovo. Both Bosnia and Kosovo turned into long-term projects for NATO and its members. Forces under NATO command remained in Bosnia until 2004, when the mission was handed over to the EU. Meanwhile, some four thousand troops under NATO command remain in Kosovo as of this writing. Indeed, the NATO operation in Kosovo is expected to last for several years to come.

The missions in Bosnia and Kosovo were by no means easy, and they were at times accused of being futile and not truly helping Bosnia and Kosovo on a path toward lasting peace. Still, NATO casualties were extremely low (and mostly due to accidents, not hostile action), and the missions must be judged as successes, especially given the downright poor results of other peacekeeping operations around the world at the time. With Bosnia and Kosovo, NATO had found its new rationale and utility in the new world order, which at the time promised to be one dominated by nasty civil wars, criminal gangs with international linkages, and state instability. In a 1999 speech the then–NATO secretary general Javier Solana clearly set out the new mission for the alliance. "The NATO of the 21st century will be a promoter of security: taking on new missions to manage crises; tackling new risks; and, perhaps most important, working with every country in the Euro-Atlantic area to build security through cooperation."[3]

After the 9/11 attacks by Al Qaeda in New York, Washington, and Pennsylvania, the strategic reorientation of NATO toward an expeditionary alliance accelerated further. But the George W. Bush administration was at first reluctant to give NATO a major role in the campaign in Afghanistan to root out Al Qaeda and the Taliban and to stabilize the country. Washington's first instinct was to rely on a loose coalition of the willing, which still included many NATO members, in order to avoid the sometimes lengthy negotiations and compromising that is a matter of course within any formal alliance. NATO's first task in Afghanistan was limited to providing security in Kabul's capital region. But as the fighting in Afghanistan dragged on and the United States became mired in a second counterinsurgency campaign in Iraq, Washington had a change of heart and sought a larger NATO role in Afghanistan. By the end of 2003, NATO took on full responsibility for the war in Afghanistan, albeit with a dominant American role. The alliance's campaign in Afghanistan would prove to be by far NATO's longest and most ambitious mission. At its height in 2012, the NATO effort would include some 130,000 troops from NATO member and partner nations (such as Sweden, Finland, Australia, and New Zealand). At the conclusion of the ISAF mission in 2014, some 1,100 NATO soldiers had been killed in Afghanistan, alongside nearly 2,300 Americans.

The peacekeeping missions in southeastern Europe and the shooting war in Afghanistan would deeply change NATO and the militaries of the member countries. Counterinsurgency and ground-centric expeditionary operations were in, and planning and equipping for major power war or significant maritime operations were definitely out. The Afghanistan campaign also drove other alliance initiatives that were seen as helpful to this type of expeditionary operation, such as the C-17 consortium, under which a number of NATO members pooled their resources to buy and operate C-17 heavy-lift aircraft that could ferry troops, equipment, and supplies to distant locations. This drive toward a small wars and ground-centric approach was perfectly understandable given the security challenges at the time and the widely held assumption, found in both Washington and in European capitals, that this was the new and enduring normal, or, in the words of several senior American military leaders at the time, it was "The Long War" that would continue for a generation or more.[4] NATO and U.S. casualties in Afghanistan amounted to nearly four thousand dead during operations in that country, and the expenditure of blood and treasure tends to focus the mind of political and military decision makers, as it indeed should. This transformation of NATO was to no small degree prodded forward by a Washington eager to get its European allies to share the burden in global crisis response. During a visit to Romania in October 2004, then–secretary of defense Donald Rumsfeld noted that if NATO could not quickly respond to transnational terrorism then "you don't have a military alliance for this century."[5]

The membership of NATO also rapidly grew in the post–Cold War period, and this also oriented the alliance toward a continental focus and mindset, which increasingly drew the alliance's collective mind away from the maritime domain, and in particular the North Atlantic. This was a radical departure for the formerly Atlantic-focused alliance. NATO's original members were all, to some degree or another, maritime nations, many of them with an Atlantic instinct. It is thus no accident that the Atlantic ended up in the name of the Cold War U.S.-European defensive pact. Indeed, all the Western nations that were or had been great maritime powers were either founding or early members of NATO, including the United States, Britain, France, Spain, Italy, Greece, and Portugal.[6] As NATO grew to incorporate the formerly communist eastern Europe, it also brought in nations that were either landlocked, such as Hungary, the Czech Republic, and Slovakia, or nations with a limited coastline but who fundamentally viewed themselves as continental nations, including Poland, Romania, and the Baltic states. Sea blindness, or the inability to appreciate the centrality of the maritime domain in terms of trade, security, and connections to the larger world, is something that even maritime nations must contend with, but it is understandably even more present in nations that are not directly connected to the sea or that lack a tradition of raising and maintaining navies or merchant fleets. Indeed, during the period of Western triumphalism some even proposed to orient NATO even further away from its Atlantic roots by creating a "global NATO" with, one day, members hailing from Latin America and the Asia-Pacific region, all contributing to global security under the umbrella of NATO.[7]

A decade into the twenty-first century, the new era of NATO activism far away from the North Atlantic region was codified in the alliance's strategy. NATO's new Strategic Concept, agreed to by all of its members in 2010, confirmed this new orientation by stating that "[t]oday, the Euro-Atlantic area is at peace and the threat of a conventional attack against NATO territory is low. That is an historic success for the policies of robust defence, Euro-Atlantic integration and active partnership that have guided NATO for more than half a century." The concept went on to explain that perhaps the risk of state-on-state war remained in ways that would affect NATO and its members, but those conflicts would almost surely happen far away from the territory of the alliance.[8]

NATO's Dwindling Navies

During this period of NATO activism, first in Europe's periphery and then later on the other side of the world, navies played second fiddle to the armies and air forces of the alliance, with the latter two branches often finding themselves engaged in combat, a state of affairs actually never experienced by NATO and rarely by most of its members during the Cold War. But the ships and submarines of NATO's members were by no means lying idle at the pier. NATO member navies played supporting roles during the operations in the Balkans and were later sent to patrol the Mediterranean for terrorists and smugglers in the wake of 9/11 under Operation Active Endeavour, NATO's first Article V (the defense guarantee clause in NATO's founding charter) effort in the alliance's history. As the threat of piracy emerged off the Horn of Africa in 2008, ships sailed under a NATO flag, alongside a similar EU mission and a U.S.-led coalition of the willing, to suppress the Somali pirates and escort humanitarian food shipments to Somalia. Still, these efforts were essentially sideshows or supporting actions that did not garner much public attention in comparison to NATO's high-profile but also grinding, frustrating, and casualty-generating effort against the Taliban and Al Qaeda in Afghanistan.

Out of these relatively modest maritime missions grew a sense across NATO that western navies were just not a good fit for the new security environment. It seemed in some ways unsatisfying to send high-end warships to hunt terrorists, look for smugglers, or suppress pirates. Indeed, to some it seemed that private security companies, organized along the same lines as those operating in Iraq and Afghanistan, were a better antidote to nonstate maritime security challenges than the navies of the West.[9] Besides, at this point NATO lacked a coherent strategy for its maritime forces and what their appropriate role would be in a twenty-first century dominated by nonstate threats and rogue state spoilers who were more of a nuisance to the international shipping lanes and global trade rather than a more direct threat to the physical security of the NATO members themselves.

NATO's Allied Command Transformation (ACT) in Norfolk, Virginia, the alliance's developer of new concepts and doctrines and in-house think tank of sorts, began a new effort in 2009 that took on NATO's role in ensuring access for itself and for legitimate commercial activity in the so-called "global commons" (the air, space, sea, and cyber arenas). In the maritime-focused chapter of the resulting report on NATO and the global commons, ACT pointed to how all NATO members are reliant on free and unfettered global trade and energy flows, which could be threatened by, among other things, piracy and terror networks. The report therefore suggested that NATO's maritime forces could serve as a provider of security in the maritime commons, along with acting as an example of good behavior at sea for the growing naval powers in the Asia-Pacific region. The maritime domain and naval forces were also viewed by ACT as the best conduit for NATO to maintain the partnerships with countries such as the United Arab Emirates, Australia, and Japan, relationships originally forged during the campaign in Afghanistan.[10]

While ACT's thinking on the maritime domain and NATO's role there mirrored much of the thinking across the alliance at the time, the ultimate reflection and codification of this new reality for NATO's role at sea and European navies can be found in NATO's own Alliance Maritime Strategy, which was released with little fanfare in 2011. It replaced the NATO maritime strategy, which was first formulated in 1984, at a time of heightened tensions with the Soviet Union in the late Cold War. As a strategy it was not a bad document in principle; it outlined threats, opportunities, resources, and a priority list for naval forces under the NATO umbrella. But the threats and challenges in the 2011 strategy were far different than the ones outlined in 1984. The emphasis was now clearly on nonstate and unconventional threats. For example, the Alliance Maritime Strategy had this to say about the security challenges at sea facing the transatlantic alliance:

> The oceans connect nations globally through an interde-pendent network of economic, financial, social and political relationships. . . . The maritime environment includes trade routes, choke points, ports, and other infrastructure such as pipelines, oil and natural gas platforms and trans-oceanic telecommunications cables. . . . Global trade relies upon secure and low-cost international maritime transportation and dis-tribution networks that are vulnerable to disruption, to the extent that even short interruptions would seriously impact international trade and Allies' economies. . . . The maintenance of the freedom of navigation, sea-based trade routes, critical infrastructure, energy flows, protection of marine resources and environmental safety are all in Allies' security interests."[11]

The above description of the strategic environment in the mari-time domain is a fine one, and it lays out the many things that navies and nations must think about when it comes to protecting the inter-national flow of goods, resources, and communications across the oceans from terror groups, pirates, and criminals. It says nothing, however, about what is at stake in the maritime domain when the foe is a determined state opponent that seeks to challenge America's and NATO's superiority at sea. In short, NATO's own maritime strategy had essentially written off the possibility of a return to great-power rivalry and its implications for the maritime domain and the sea services. In any event, after its release in 2011 the new maritime strategy quickly sank into relative obscurity, as it was rolled out just as NATO was preparing to take on its next expeditionary operation: the air campaign to support the Libyan rebels against the crumbling Qadaffi regime in Libya, which went on to consume the alliance and the media's attention for the next three months.[12]

NATO's new Strategic Concept, which preceded the maritime strategy and got plenty of attention from decision makers, said relatively little about NATO's role in the maritime domain, but what it did say also pointed the alliance toward a future where it would help manage the global maritime domain against disruptions from pirates and criminals, not as an arena where NATO would have to deploy assets to provide deterrence against state aggression or compete with revisionist powers. The Strategic Concept even mused, like the Alliance Maritime Strategy, that NATO might one day play a more prominent role as the guarantor of the safe and secure flows of energy supplies around the world.[13] The EU's own maritime security strategy was rolled out in June 2014, and it too saw a future where Europe's reliance on the international trade lanes would be most at risk from organized crime, environmental factors, and terrorism at sea, not from great-power competition.[14] The EU is of course not a military alliance like NATO, but the EU has sought a larger role for itself in defense and security policy for nearly two decades. Its description of the current and future threats in the maritime domain very much echoed that laid out by NATO's Alliance Maritime Strategy in 2011.

National defense and maritime strategies from the period reflected the same sense that great-power conflict was a thing of the past and that armies, air forces, and navies now needed to focus on low-end threats and the policing of a world order that was largely benign and had moved beyond great-power competitions. The 2010 Strategic Defense and Security Review (SDSR) from the British government reached a similar conclusion about the global security environment and what the British military should focus on, including "key counter-terrorist capabilities . . . a transformative program for cyber security . . . focus cross-government effort on natural hazards . . . and . . . preventing international military crises." U.S. maritime strategy underwent several evolutions after the 1980s strategy, which had placed such a premium on operating in the far north Atlantic. The U.S. "Cooperative Maritime Strategy" from 2007, signed on to by the Navy, Marine Corps, and Coast Guard, begins by stating that nations "prosper because of this system of exchange among nations, yet recognize it is vulnerable to a range of disruptions that can produce cascading and harmful effects far from their sources." The strategy later proceeds to identify "terrorists and extremists; proliferators of weapons of mass destruction; pirates; traffickers in persons, drugs, and conventional weapons; and other criminals" as key threats in the maritime domain, or actors who may leverage the maritime domain for their activities. More traditional challenges, such as nations operating increasingly sophisticated submarines or regional powers seeking to extend their influence, are duly mentioned in the Cooperative Maritime Strategy, but its focus on policing the maritime domain in concert with others, perhaps even the Russians and Chinese, is abundantly clear.[15] The Cooperative Maritime Strategy was preceded by the "Thousand Ship Navy" concept in 2005, which focused on building the capacities and capabilities of friendly navies to help police the maritime domain. It too betrayed an implicit assumption that the maritime future was essentially one of cooperation, not competition and potential conflict against a state adversary at sea. These American statements about the future of security in the maritime domain also helped shape the thinking among allies across the globe, not least in Europe. Indeed, it is hard to overestimate the decisive, yet subtle, influence that American strategic thinking has had on NATO's orientation and allied concepts and strategies over time. While Washington certainly does not always get what it wants out of NATO, which is after all an institution driven by consensus, American strategic concepts, visions, and approaches have a way of working their way into NATO's political and military bloodstreams over time.

The changing global security environment, along with the new strategies and missions adopted by NATO, also began to drive a transformation of European navies. They began to slowly but surely change away from high-end warfighting and toward the threats and challenges that the strategy documents called for. This was reflected in both the type of training and exercises that European navies conducted starting in the 1990s and up to 2014. For example, the annual naval exercise BALTOPS, led by U.S. Naval Forces Europe and first held in 1971 in the Baltic Sea, quickly evolved away from its warfighting roots after the end of the Cold War. In the 1990s BALTOPS turned into a practical way to engage with the navies of the former Warsaw Pact nations, such as Poland and the Baltic states, that now sought to become NATO members. And by the 2000s BALTOPS' orientation was clearly toward maritime security tasks and supporting expeditionary operations. The 2005 iteration of the exercise even included the Russian Baltic Sea Fleet as a full exercise participant and was at the time described by U.S. European Command as an event to "improve interoperability with allies and PFP countries by conducting peace support operations at sea to include a combined amphibious landing and a scenario dealing with potential real world crisis."[16] Russian warships returned as participants in the more maritime-security-focused BALTOPS exercises until 2012.

The reorientation toward softer maritime security missions also influenced the type of warships that European navies bought. Germany's new frigate class, the *Baden Württemberg*—the first ship laid down in 2011 and brought into service in late 2016—came with little for hunting submarines, but it did have the ability to operate for extended periods of time away from its homeport and in tropical conditions, and it came with nonlethal weapons such as water cannons.[17] The Dutch navy sold off four of its eight M-frigates in the mid-2000s, and opting for a new class of ocean patrol vessels that were far larger, but also focused on the lower end of the threat spectrum and with the envisioned main role of helping to police the world's oceans.[18] At the same time, European submarine builders began to market their boats as primarily intelligence-gathering platforms or as the ideal vehicles for delivering special forces to distant beaches in a covert fashion instead of, perhaps, the most effective way to hold far larger surface fleets at risk or to detect, track, and fight other submarines.

This reorientation was further accelerated by the defense austerity imposed by governments across the continent in the wake of the Eurozone crisis, which was triggered by the financial crisis in the United States in 2008. To be sure, European defense spending had been on the decline since the end of the Cold War, shrinking by nearly 20 percent between 1990 and the mid-2000s, a reduction made more stark in light of the fact that it occurred during a period when European GDP rose by nearly 55 percent. But the austerity imposed with the coming of the Euro crisis meant another round of pronounced and deep cuts.

The United Kingdom was especially hard hit by the recession brought on by the popping of the American real estate bubble in 2008 and the subsequent Eurozone crisis in Europe. The British government applied fiscal austerity across the board, and the British armed forces took their fair share of the cuts; but the austerity struck the Royal Navy in particular. The 2010 SDSR decided to cut the number of destroyers and frigates in the Royal Navy from twenty-three to nineteen, with the option to reduce the number further to only twelve. At the same time, the SDSR rushed the British aircraft carriers HMS *Ark Royal* and the HMS *Illustrious* into early retirement, leaving a gap in British carrier aviation that will last nearly a decade until the new HMS *Queen Elizabeth* comes into active service in 2020.[19]

Meanwhile, Germany reduced defense spending by nearly 22 percent between 2011 and 2014, which included a cut in its planned procurement of new corvettes from fifteen to five.[20] Italy and Spain introduced additional cuts of 9 and 13 percent, respectively, and the Netherlands planned a 13 percent reduction until 2015.[21] And while all services experienced cuts, European ground forces were comparatively spared, given that they were engaged in protracted and bloody operations in places such as Iraq and Afghanistan. Navies, on the other hand, unable to point to immediate and urgent relevance to the operations of the day, faced deep cuts. That dynamic was especially true for what were and are the sharpest tools in the naval drawer against advanced and modern submarines: frigates, submarines, and MPAs.

During the last phase of the Cold War, NATO's navies had some 230 frigates at their disposal for a range of tasks, one of the most important ones being ASW. By 2013 that number had declined to fewer than 100.[22] The falling frigate numbers were especially pronounced among some of the leading European navies. Between 1995 and 2013 the Royal Navy went from 23 to only 13 frigates. During the same period France cut its frigate force from 35 to only 18. Meanwhile, Spain went from 18 to only 6 frigates.[23]

The story of European MPA fleets in the post–Cold War era in particular is one largely of decline and being pressed into service in roles far from what they were originally intended for. And this decline was further hastened by the deep defense cuts after 2008. The Netherlands, one of Europe's premier maritime nations, which had built its empire and wealth by leveraging the sea, gave up its entire fleet of P-3 Orion aircraft in the year 2000 and walked away from the long-range MPA mission, seeing little use for the fixed-wing MPA fleet in the new security environment. Some of the Dutch P-3s were snapped up by Germany, which saw an opportunity to gain lightly used but well-maintained and recently modernized MPAs. But this move in itself meant that Germany could postpone a full recapitalization of its MPA fleet, which had been on the books in the early 2000s. Meanwhile, France decided to modernize only half of its fleet of Atlantique 2s, leaving the rest with outdated systems and aging airframes. In southern Europe, Greece mothballed its own fleet of P-3s in 2005, and they remained in this state for more than a decade.

The most notorious example of the decline of airborne ASW in Europe was the British decision in 2010 to scrap its in-development MPA, the MRA4 Nimrod. Along with the early retirement of a carrier, the striking of four frigates from the Royal Navy's rolls, the 2010 SDSR effectively withdrew the United Kingdom from long-range ASW in the North Atlantic. The 2010 SDSR described a world in which the British armed forces would have to face challenging nonstate threats and asymmetric actors, but where the chances of facing a peer competitor in high-intensity conflict were nearly zero. It also made the strategic bet that the world would remain one of lower-end threats in the coming decades.[24] Along with a changing security environment and the need to cut defense spending in response to national austerity, the decision to get out of MPAs was in no small part driven by diverging service interests. Britain's MPA fleet was operated by the Royal Air Force, not the Royal Navy, and when faced with the need to cut capabilities, the Royal Air Force's instinct naturally drove the service to preserve, to the extent possible, what was seen as core air power components, such as fighters, lift, and reconnaissance planes, instead of the Nimrod. At the same time, the Royal Navy had little appetite to further cut its ship numbers to help preserve a British MPA capability.

All in all, European MPA fleets lost nearly half of their complement during the decade and a half between 2000 and 2016.[25] But the ones that were left were worked hard. The counterpiracy mission off the Horn of Africa, which for a time saw the involvement of the United States, NATO, and the EU, along with ships from India, China, and Russia, used MPAs to monitor the vast maritime space for pirates that could cover large distances by using so-called motherships that would send out smaller and fast skiffs when a promising cargo ship was spotted. MPAs were also used in Afghanistan for reconnaissance missions and in the Mediterranean under Operation Active Endeavour. Before their retirement from service and sale to Germany, Dutch P-3s saw service in the Caribbean to support the hunt for drug smugglers using speedboats to ferry their illicit cargoes between islands. MPAs also took part in the NATO mission over Libya in 2011, where a U.S. Navy P-3 fired a Maverick air-to-ground missile against a Libyan patrol boat, which put it out of action. A nation must of course use all of its military resources that it has on hand when responding to a real-world operation, but it does not come without costs in terms of lower readiness for other operations or deterioration in other skill sets. Airborne ASW is a unique undertaking that requires methodical and consistent rehearsal and exercises, preferably together with other units under realistic conditions. And while it is and has always been technology-intensive, at the end of the day a well-trained and experienced crew makes all the difference. All those hours, days, and months spent hunting for Somali pirates or providing overhead ISR for coalition operations in Afghanistan meant time not spent honing the craft of airborne ASW.[26]

European submarine fleets experienced a similar decline in the decades after the Cold War. In the year 2000 the European nations operating submarines in the broader North Atlantic included the United Kingdom, Norway, France, the Netherlands, and Denmark. Together, they operated some fifty nuclear and conventional attack submarines. By 2016 that number had shrunk to only a little more than half, or twenty-eight boats.[27] Along with the reduction of active submarines in European fleets, new classes of submarines were either delayed or scrapped altogether. For example, in the mid-1990s, Sweden, Denmark, and Norway joined forces to design and produce a new class of submarine (the *Viking*), which could serve the needs of the Nordic navies that operated both in the littorals of the Baltic Sea and the deep waters of the Norwegian Sea and the Atlantic. All three nations were in need of a new class of submarine, but the project fell apart in the early 2000s. Sweden and Norway have not taken on a new submarine class since then, and Denmark withdrew from operating submarines in 2004.[28] During the same period Germany decommissioned its *206*-class submarines early and reduced its planned buy of *212* submarines by two boats.[29]

And like the shrinking fleet of MPAs, the remaining submarines were pressed into service in missions where their power and capabilities made them an awkward fit. Denmark dispatched a submarine to support the U.S.-led invasion of Iraq in 2003, while Norway contributed a submarine to the NATO-led Operation Active Endeavour in the Mediterranean, the allied effort to monitor the sea for terrorism and escort at-risk commercial shipping. The Netherlands also sent one of its Walrus boats all the way to the Horn of Africa to help combat piracy off the east coast of Africa. In the early 1990s Canadian submarines were detailed to help track not only

A twenty-four-ship convoy off Newfoundland heading across the North Atlantic in 1942. Convoys such as these played a pivotal role in keeping Britain and the Soviet Union in the war against Germany and also carried U.S. troops, vehicles, equipment, and supplies for the U.S. war effort in Europe. *Naval History and Heritage Command*

The German submarine *U-515* under attack in the North Atlantic by U.S. Navy ships and aircraft in April 1944. The Battle of the Atlantic was the longest continuous campaign of World War II, running from the fall of 1939 to the surrender of Germany in early May 1945. *Naval History and Heritage Command*

A Soviet *Golf II* submarine on the surface while exiting the Baltic Sea with the frigate USS *Pharris* from the Standing Naval Force Atlantic observing. Taken during the summer of 1978, this photo highlights how closely NATO and Soviet naval forces would at times operate. *Naval History and Heritage Command*

Artist's rendering of a Soviet SSBN base on the Kola Peninsula during the late Cold War. The Kola played a key role in Soviet nuclear deterrence as a staging area for the submarine-based nuclear deterrent, but it was also seen as vulnerable to outside intervention by the United States and NATO. *Defense Intelligence Agency*

The Dutch frigate *Tromp* and a U.S. ASW helicopter during Exercise North Star in 1991, which proved to be one of the last major naval exercises in the North Atlantic as the Cold War came to its abrupt end. *Joseph Dorey/U.S. Navy*

NATO member forces practice natural disaster response in Iceland in 2002. After the end of the Cold War these types of humanitarian exercises replaced the training events focused on high-end maritime war fighting in the North Atlantic. *JO2 Stephen Sheed/U.S. Navy*

A Dutch frigate, part of NATO's operation Ocean Shield, intercepts a pirate dhow in the Indian Ocean in 2012. NATO naval forces found themselves far from their traditional environments and missions in the post–Cold War world. *NATO*

A Russian submarine from the Northern Fleet awaiting dismantlement at the Sevmash yard on the Kola Peninsula. The dismantlement of the Northern and Pacific Fleets' nuclear submarines was an effort of nearly epic proportions in the post–Cold War period. *Defense Threat Reduction Agency (DTRA)*

Severodvinsk, the first boat in Russia's new *Severodvinsk* multipurpose SSN class. First laid down in 1993, it joined the Northern Fleet in 2013. The *Severodvinsk* class approaches that of U.S. and allied submarines in terms of quieting and sensors and can carry up to forty Kalibr cruise missiles. *Russian Ministry of Defense*

The improved *Kilo*-class *Rostov-na-Donu* on its way to join Russia's Black Sea Fleet in late October 2015, having finished up sea trials and cruise-missile tests in the Barents Sea with the Northern Fleet. On December 8, 2015, the *Rostov-na-Donu* struck targets in Syria from the Mediterranean with Kalibr cruise missiles, marking the first time the Russian navy fired cruise missiles from submarines in anger. *Russian Ministry of Defense*

A submarine-launched land attack version of the Kalibr cruise missile on display. The Kalibr has been used extensively by the Russian navy in strikes against targets in Syria. Its long range means that it can be fired at targets across northern Europe from close to Russian naval bases on the Kola Peninsula. *Vitaly Kuzmin/Creative Commons*

In 2017, U.S. MPAs were once again flying out of Keflavik, Iceland, for ASW training and exercises. Depicted is a U.S. Navy P-8 Poseidon in late April of that year. Note the weather conditions in the far North Atlantic, in spite of the spring season. *Lt. (jg) Matthew Skoglund/U.S. Navy*

Dynamic Mongoose, a deep-water ASW exercise off the coast of Norway in 2015. By that year NATO member navies were once again drilling for antisubmarine warfare in the North Atlantic. *NATO Maritime Command*

Close encounters between Russian and U.S. and NATO member forces in the maritime domain are back. In this 2016 incident, two Russian Su-24 jets repeatedly pass USS *Donald Cook* at close range while the *Donald Cook* operates in the Baltic Sea. *U.S. Navy*

An overhead view of satellite receivers on the Svalbard islands. The archipelago close to the top of the world has gained additional strategic importance in the age of space-enabled operations. *Erlend Bjortvedt/Creative Commons*

Unmanned platforms such as the *Sea Hunter* currently being developed by DARPA may provide the United States and its NATO allies with a future edge in the contest with Russia's increasingly quiet and sophisticated submarines in the North Atlantic. *John F. Williams/U.S. Navy*

drug smugglers, but even foreign vessels fishing illegally in Canadian waters. The submarines that deployed on these many and varied missions undoubtedly contributed to them through intelligence gathering and the ability to covertly monitor the shipping lanes and the broader maritime domain.[30] But it was lost on few in the broader defense community that it was suboptimal to send such powerful undersea platforms, capable of sinking capital ships and hunting down other submarines, to take on adversaries armed with little more than assault rifles and rocket-propelled grenades.

To compare ship and aircraft numbers and their aggregate capabilities across time is an inexact undertaking in the best of times. It is true that newer classes of ships are often much more capable than their predecessors, meaning that a drop in numbers can to some extent be offset by the enhanced capabilities found in the new ship class. For example, a new frigate may have a new sensor suite and extended-range weapons, allowing it to reach further out in the battlespace, track multiple targets simultaneously, and be more robust in the face of, say, electronic warfare. A new class may also be of a flexible design, meaning it can swing between roles, such as ASW, mine hunting, or air defense with the use of various shipboard modules. But this can only be taken so far; a single ship can only be at one place at any given time, and it is still subject to training and maintenance periods, which will make it unavailable for operations.[31] NATO's aggregated naval power may have become more sophisticated during the post–Cold War period and more attuned to tackling maritime security tasks far from the North Atlantic region, but the drastic cuts meant that there were simply fewer submarines, frigates, and MPAs to patrol and be present in the huge North Atlantic and elsewhere.

But the falling number of frigates, submarines, and MPAs in the European inventory was not the only consequence of the reductions in defense spending across the continent. Readiness and the operational availability of naval units also declined. Germany was unable to send any of its remaining submarines to sea for a time, due to the lack of spare parts.[32]

The period between 2000 and 2015 did not see the U.S. Navy
under pressure from lack of funds in the same way as its European
counterparts. The U.S. Navy did, however, experience a dramatically
increased tempo due to the wars in Iraq and Afghanistan. And the
changing security environment meant that the U.S. Navy saw a similar turn away from ASW and high-end naval warfighting. Thus the
U.S. Navy also lost much of its instinct for hunting submarines after
the end of the Cold War. In an age dominated by U.S. interventions in
fragile regions, the U.S. Navy increasingly played a supporting role,
and little suggested a return to great-power competition and high-
end naval combat. This period left telltale marks in the U.S. ASW
arsenal, just as among America's NATO allies. For example, U.S.
aircraft carriers have carried ASW aircraft since shortly after World
War II, where they provided an extended ASW screen for the otherwise vulnerable carriers. In response to the increasing threat from
Soviet submarines during the late Cold War, the U.S. Navy developed
the S-3 Viking, a jet with a magnetic anomaly detection system, radar
detection equipment, and the ability to deploy up to sixty sonobuoys.
The further refined S-3B Viking was introduced in 1987, nearly too
late to have a role during Cold War ASW efforts. After the Cold War
the U.S. Navy found new roles for the S-3 Viking jet, including as
an onboard tanker for carrier-based aircraft, which meant that it
was no longer dedicated to its original ASW mission.[33] The last S-3B
was retired from active service in 2009, leaving U.S. aircraft carriers without an onboard ASW platform other than the short-range
SH-60 Seahawk helicopters they carry for a range of missions. And
just like its European counterparts, U.S. Navy MPAs were put into
service conducting, among other things, overland ISR and monitoring the sea-lanes for pirates and smugglers.[34] And as was the case in
Europe, U.S. MPA numbers fell precipitously during the post–Cold
War period, from twenty-four squadrons down to twelve.[35]

The poor state of the U.S. Navy's capacity for ASW was showcased
by a unique exchange arrangement between the United States and
Sweden, which received little fanfare or notice when first started
in 2005. This cooperative effort between the United States and

Swedish navies included the transfer of the Swedish diesel submarine the HMS *Gotland* to San Diego for a one-year period, where the submarine and its crew would act as the opposing force during naval exercises for ships heading out for deployments in the Asia-Pacific region. The *Gotland* was the first in-service class of submarines in the world using air independent propulsion machinery, which meant that she could remain submerged for far longer than a regular diesel submarine, which would need to come to the surface or snorkel depth every twenty-four to forty-eight hours to run the diesels and recharge the batteries needed for subsurface propulsion. During the first year ported in San Diego the HMS *Gotland* spent some 160 days at sea and racked up surprising results against the U.S. Navy's surface ships and ASW aircraft. While public reporting about the outcomes of the ASW exercises between the U.S. Navy and the *Gotland* is limited, it seems clear that the *Gotland*, among other things, achieved an advantageous attack position versus the USS *Ronald Reagan* during a Joint Task Force exercise in late 2005. Norman Polmar, one of the most distinguished American naval experts, told a local San Diego TV station at the time that the *Gotland* "is running circles around our carrier battle groups."[36] With the results in hand on what a modern conventional submarine could achieve against U.S. naval forces that had not focused on ASW since the Cold War, the U.S. Navy quickly negotiated a one-year extension to the *Gotland*'s stay in San Diego for further exercises and evaluation.[37] The ASW-cooperation initiative between Sweden and the United States was clearly aimed at emerging threats in the broader Asia-Pacific region, but the initially poor results on the side of the U.S. Navy were just as applicable to contingencies in the North Atlantic. But even among American advocates of a renewed U.S. focus on ASW, the threat was not seen as coming from near-peer competitors with advanced nuclear submarines. Instead, the future threat was thought of as primarily coming from rogue states, such as Iran and North Korea, with relatively small and conventional submarines that operate in the littorals.

But naval power also depends on far more mundane matters such as the training and exercises that crews have been able to put in, maintenance of the ships, or, for that matter, the availability of weapons to use in case of a conflict. For example, the U.S. Navy's mainstay ASW torpedo for use by submarines, the Mark-48, was not purchased in any great numbers beginning in the year 2000, and stocks are drawn down over time, even during peacetime, through exercise shots or decaying components, which forces the removal of the torpedo from the inventory.[38] The then–chief of naval operations, Adm. Jonathan Greenert, told the U.S. House of Representatives in 2015 that U.S. Navy torpedo stocks were some 30 percent below wartime requirements.[39] Nations tend to be reluctant to publicly report the status of their stocks of missiles, torpedoes, and other weapons. This is true for even Western militaries that otherwise are relatively transparent about their defense planning and military capabilities. But if experience is any guide there is little to suggest that European navies have done much better maintaining their stocks of ASW weapons. And in war, munitions tend to be expended fast and likely at higher rates than originally predicted. During the British pursuit of the Argentinian submarine *San Luis* during the Falklands War, the Royal Navy expended some fifty ASW torpedoes without achieving a kill of the submarine (for more detail see chapter 2). Indeed, other types of weapons have already been proven to be in short supply among European militaries when the need for them has arisen. In both NATO's Operation Allied Force over Serbia in 1999 and Operation Unified Protector over Libya in 2011, European air forces ran dangerously low on precision-guided bombs. The crisis was only averted by the United States rapidly shipping more bombs to its allies or by a quick procurement of munitions from third nations who had some to spare.[40] Nothing suggests that the situation is any different for ASW weapons and sensors such as sonobuoys that can be deployed in great numbers by MPAs.

Taking Apart the North Atlantic Network
By the end of the Cold War the United States and its European allies around the North Atlantic had built an impressive, complex, and broad network of bases, infrastructure, and sensors to monitor and

operate across the North Atlantic. It too would face radical changes, indeed disassembly, during the post–Cold War period. NATO's maritime command structure changed radically in response to the end of the Cold War and a much more active alliance on the world stage. ACLANT, the core part of NATO's command structure for the North Atlantic during the Cold War, was determined to be no longer vital due to the lack of a threat against shipping across the Atlantic. This was in no small part driven by the United States, which in the wake of the 9/11 attacks was reorganizing its own command structure; and it signaled to its NATO allies that America was no longer particularly interested in being responsible for defending the North Atlantic maritime domain against what appeared then to be a nonexistent enemy.[41] But ACLANT was not exactly retired. NATO still saw the need for an alliance presence in North America (since virtually all the rest of NATO's commands, agencies, and schools were located in Europe) and created out of ACLANT the aforementioned ACT, which was charged with developing future concepts and education and training to help drive NATO's transformation in the post–Cold War era. The maritime functions previously held at ACLANT were transferred to NATO's new centralized Maritime Command (MARCOM) in Northwood in the United Kingdom. The lack of a regionally oriented maritime command structure on both sides of the Atlantic also meant that it was harder to pull together naval forces for multinational exercises in those areas. Regional maritime basing across the North Atlantic region was also closed or reduced in the post–Cold War period, being seen as less relevant to the new security environment.

During the Cold War, Norway maintained a hardened submarine base by the name of Olavsvern in northern Norway, within easy reach of the Barents Sea and the Kola Peninsula. The base included hardened submarine pens that were nearly impervious to long-range strikes, including from nuclear weapons, which allowed for submarines to be supplied and maintained in relative safety. Opened in 1967, Olavsvern represented NATO's northernmost submarine base in a critical region. The base was closed in 2002 and sold to a private investor on

Norway's equivalent of eBay in 2009. The new owner later leased the base to a Russian company that provides services to Russia's oil and gas conglomerate Gazprom. Since the lease began, Russian research vessels, which play a role in conducting underwater seismic surveys for military purposes, have begun to call at Olavsvern.[42] The British RAF base Kinloss, in the north of Scotland, which had served as the UK's main hub for MPA operations in the Atlantic during the Cold War, was ordered closed in 2010 as a direct result of the British decision to walk away from operating MPAs earlier that year. The Dutch naval air station at Valkenburg, which provided an excellent perch for MPAs patrolling the North Sea and areas north of the GIUK gap, was shuttered in 2006, as the Netherlands decided to leave the MPA business and get rid of its P-3 Orion fleet. Meanwhile, on the other side of the North Atlantic, the 1990s saw Canada shuttering most of the onshore infrastructure in Nova Scotia that had supported both Canadian and U.S. operations in the North Atlantic during the Cold War, including shore stations supporting SOSUS, radio stations, and air bases that had previously hosted MPAs and ASW helicopters.

The United States also gave up its basing on Iceland at Keflavik. The size of the base had begun to shrink nearly immediately after the end of the Cold War, to the point that by 2005 only four F-15s were permanently stationed there, along with its ground crew, air traffic controllers, and a contingent of search-and-rescue helicopters—in case one of the F-15s ever crashed and the pilot needed to be plucked out of the North Atlantic. In early 2006 then–secretary of defense Donald Rumsfeld, who had made himself a name as a secretary eager to transform the U.S. military out of its Cold War posture, pushed to have Keflavik completely closed, while noting that the base had cost U.S. taxpayers some $2 billion since the end of the Cold War and had an annual bill of some $260 million for the four F-15s there.[43] The U.S. pullout from Keflavik broke records in terms of its speed. The last American service member left Keflavik some six months after Rumsfeld made the decision at the Pentagon. The departure caused a minicrisis in the U.S.-Icelandic relationship, as Washington and Reykjavik negotiated the closure procedure and how the

U.S.-Icelandic defense relationship could be continued without a permanent U.S. presence. The Icelandic surprise at the U.S. decision to depart was sincere, but the anger was also fueled by the loss of a key lifeline for Icelandic fishermen. When their fishing vessels went under in the unforgiving North Atlantic they too were picked up by the U.S. helicopters stationed with the F-15s at Keflavik.[44] Along with the closing of Keflavik, the manning of the U.S. airfield on the Azores in the southern North Atlantic was also drastically reduced, going from a Cold War high of two thousand to only two hundred. And in 2015 serious discussions were being held at the Pentagon to completely withdraw the U.S. presence from the Azores.[45]

All in all the U.S. Navy's presence in Europe was cut in half by the mid-2000s compared to late Cold War numbers, with personnel strength going from roughly 14,000 to only 7,500. Most of the ships homeported in Europe were also returned to naval bases in the United States, leaving a handful of destroyers for the missile defense mission against Iran's growing ballistic missile capabilities, and the command ship the USS *Mount Whitney*.[46] This development largely mirrored that of the U.S. Army and Air Force, which were also closing bases and returning equipment and personnel to the United States or bringing them to bases closer to the big counterinsurgency fights in Iraq and Afghanistan. What was left of the U.S. Navy in Europe was also swung toward the south and the Mediterranean and away from the North Atlantic. In 2005 U.S. Naval Force Europe closed its headquarters in London and moved it to Naples, where the commanders and the staffs would be closer to the action in the Middle East.[47] The U.S. pivot to Asia also informed and drove the U.S. Navy's global posture—with a slow but unmistakable shift of naval resources, including submarines—from the Atlantic to the Pacific, a region that was increasingly seen as the only one that would be contested and competitive in the twenty-first century. The U.S. Second Fleet, responsible for North Atlantic operations and keeping the sea lines of communication between North America and Europe open during the Cold War, was also disestablished in 2010 in an effort to shrink what was seen as unnecessary force structure and to achieve savings. Since the end of the Cold War, the Second Fleet had been engaged in maritime security operations and humanitarian efforts in, among other places, the Caribbean. These were useful efforts, but hardly the type of military operations that warranted their own fleet.

Finally, the sensor networks used to monitor the North Atlantic and its adjacent seas also fell away from attention in the post–Cold War period. Shortly after the Cold War, the SOSUS network that covered the North Atlantic, which the United States had invested so heavily in since the 1960s, began to deteriorate, with the U.S. Navy unwilling to continue footing the bill for a system that seemed to have outlived its usefulness. One by one the SOSUS terminals ashore supporting the Atlantic theater began to be turned off, with the last one in Newfoundland, Canada, and in Wales in 1995. The remaining ashore station, at Keflavik in Iceland, was closed in 1996.[48] Shutting down SOSUS was far more elaborate than simply walking away from the arrays on the seabed and turning off the lights in the onshore installations. Oftentimes it would involve physically destroying the connection between the array and the station ashore, along with burying the shore-to-sea connection in concrete to prevent access. And the budget allocation to maintain the remaining parts of SOSUS began to rapidly decline in the 1990s, from roughly $300 million in 1991 to only $60 million in 1995.[49] But SOSUS' service was not quite over yet. The scientific community along with the National Oceanic and Atmospheric Administration lobbied the Clinton White House hard to allow access to the previously classified network so that the still active part of it could be used for scientific purposes, such as studying the movement of whales across the oceans or tracking undersea quakes. Parts of the array were also used to monitor nations sticking to a treaty banning underwater nuclear tests.

In aggregate, twenty-five years after the end of the Cold War the naval picture in the North Atlantic had changed radically. European navies were smaller than ever before, but primed and ready to take on missions and threats far from their European waters. The U.S. naval presence in Europe had been more than halved and reoriented toward the Mediterranean and the turbulent Middle East. Sensor networks, basing infrastructure, and command structures intended for the Atlantic and Europe's north had been reduced or scrapped during the same period. And ASW training was far from the minds of Western navies. Then, in 2014, great-power competition and the specter of future war returned to Europe and the North Atlantic.

PART III

Contest

AFTER NEARLY THREE DECADES of quiet and cooperation in the North Atlantic domain, the security environment changed once again in 2014 with the return of great-power competition in Europe. This development came as a strategic surprise to many, not least NATO and the United States, which had been focused far away from the North Atlantic and maritime operations. At the same time, Russia began to show off new capabilities that will heavily influence the emerging contest over the North Atlantic. New actors, some of them from far away from the Atlantic, along with new technologies and a more competitive global security environment, will further tax the United States and its allies as they seek to secure the North Atlantic in the twenty-first century.

CHAPTER 10

The Return of Competition
in the North Atlantic

A Revanchist Russia

IN 2014 RUSSIA SHOOK the very fundamentals of European security, and even threw the U.S.-led global order into question, by its annexation of the Crimean Peninsula and the insertion of Russian armed forces into eastern Ukraine. The pace of events was rapid. In late November 2013 Ukrainian president Viktor Yanukovych, a leader with pro-Russian leanings, halted the implementation of an association agreement with the EU. He was pressured to do so by a Kremlin fearful that Ukraine was well on its way to leaving Russia's orbit and joining the European and transatlantic community through eventual EU and NATO membership. The Ukrainian government's move to walk away from the association agreement with the EU sparked widespread protests among Ukrainians who felt that their future lay with Europe rather than with Russia. After months of noisy and sometimes violent protests on the streets of Kiev and elsewhere in Ukraine, Yanukovych was ousted and fled to Russia on February 22, 2014. This triggered a Russian response to halt Ukraine's move toward the West, which took the form of a takeover of Ukrainian military installations on the Crimean Peninsula by soldiers bearing no national insignia. Dubbed "little green men" in the West, they were clearly members of Russia's special forces given their equipment, uniforms, and professional behavior. As part of the Russian takeover, an aged Russian cruiser was mysteriously sunk in the port outlet at Sevastopol, effectively blocking the Ukrainian navy from gaining access to the Black Sea.[1] The conflict spread to the Donbass region of Ukraine, where the Ukrainian army and volunteers battled separatists and regular Russian army units. The Ukraine crisis continues as of this writing and includes fighting across the spectrum, from disinformation efforts and cyber attacks to what is essentially trench warfare along the so-called line of contact in eastern Ukraine.

The seemingly sudden conflict between Russia and Ukraine had a stunning effect on NATO and parts of the Washington national security community. A new pattern of Russian behavior in Europe had been established, and with the move into Ukraine, Russia under Vladimir Putin had shattered the idea that war between states had been banished from the European continent, that Russia would ultimately join, however slowly and fitfully, the community of prosperous and cooperating democracies of Europe, and that great-power competition in Europe and elsewhere was a thing of the past. It began to dawn on U.S. and European decision makers that Putin also seemed bent on seeking out fissures within NATO and posturing Russia where it would be able to break NATO's cohesion, thereby ending the alliance, and on ousting the United States from its role as the key guarantor of peace and stability on the continent, a role held by the United States since the end of World War II. Achieving this would leave Russia free to dictate the terms of the relationship with its neighbors in the post-Soviet space, and perhaps with other significant parts of Europe, and knock America off the commanding heights as the leader of the global order. But the Ukraine operation was not Moscow's first military adventure since the Cold War. Russia also fought a short and sharp war with America and Western-leaning Georgia in 2008 over the disputed South Ossetia region.

The Russian war with Georgia should have been interpreted as a Russian attempt to assert itself against a U.S.- and NATO-supported nation, whose armed forces had been extensively trained and equipped by the United States to participate in NATO and coalition operations in Iraq and Afghanistan. At the time, however, most policymakers in the West were willing to look the other way, choosing instead to assume that it was a one-time event that sprang out of a particularly complicated relationship between Georgia and Russia, compounded by the incompetence of the Georgian government to effectively manage the crisis over South Ossetia. Besides, at the time the United States was in the middle of a contentious presidential election campaign between Barack Obama and John McCain, the American military was mired in Iraq, while NATO was waist deep in the protracted and frustrating counterinsurgency effort in Afghanistan.

But the 2014 Ukraine crisis would have a more lasting effect and would also lead to heightened tensions and friction between Russia and NATO, with Moscow running short-notice exercises near the territory of NATO's eastern members and stepping up its activity in the air domain with bombers and fighters that skirted, and sometimes infringed, on the national air space of NATO members. NATO's early response to the new security situation in Europe included increased ground exercises in NATO's most vulnerable states, including Estonia, Latvia, Lithuania, and Poland, along with beefed-up air policing missions along NATO's eastern flank above the alliance's smallest members that did not have their own air forces. In 2017 this would be followed by forward-based rotational battalions in Estonia, Latvia, Lithuania, and Poland, manned by units from across NATO and led by the United Kingdom, Canada, Germany, and the United States, respectively. A similar effort to boost NATO's forward presence in some of the Alliance's Black Sea members (Romania and Bulgaria) also got under way. But what was less apparent at first to senior leaders at NATO and in allied capitals was that a pattern was emerging in which much of the friction between the alliance and Russia occurred at sea, and that Russian maritime forces played a prominent role in testing and prodding NATO and advancing Russian interests in and around Europe. It was also not immediately apparent to NATO's leaders that Russian

naval power and the North Atlantic would once again play a key role
if the new nightmare scenario, an all-out war between NATO and
Russia, ever came to pass.

The Russian Navy Sails Again

In 2014 alone there were more than twenty-seven incidents or close
encounters between Russian warships and aircraft in the maritime
domain in northern Europe, including in the Baltic Sea, the North
Sea, and the Atlantic. This included an uncomfortably close pass
between a Russian fighter jet and a commercial airliner coming out
of Copenhagen airport, and Russian frigates on a missile-firing drill
entering Lithuania's economic exclusive zone, which caused real
disruption to shipping in the eastern Baltic Sea. This period also saw
the reporting of likely Russian submarine encounters in Swedish
and Finnish waters and later reporting from the French of a track of
a Russian SSBN in the Bay of Biscay.[2]

While Russia's actions at sea grew more audacious and assertive
after the Ukraine crisis, it is important to note that Russia's use of its
growing maritime muscle actually predates 2014; call it a quiet period
of operations. Warships from the Northern Fleet left home waters
for presence patrols in the Mediterranean, the Caribbean, the South
Atlantic, and the Indian Ocean in 2008 and 2009. Russian warships
also operated close to Danish and Norwegian exclusive economic
zones during the same period. At the same time, the operating limits
of the Northern Fleet's submarines were extended and under-ice
training for the submarine force was resumed. In 2009 two Northern
Fleet *Akula*-class SSNs were spotted some two hundred nautical
miles from the U.S. east coast, a first in over a decade. A year later,
in the late summer of 2010, an *Akula* boat was detected seeking to
track a British *Vanguard*-class SSBN departing its base in Scotland.
And in 2012 an *Akula* submarine and a *Sierra-II* SSN were detected
separately operating off the U.S. east coast and in the Gulf of Mexico.[3]

Russian air activity in the North Atlantic maritime domain with
aircraft flying from the Kola Peninsula also began to increase before
2014. In the first two years after the United States departed Keflavik

airbase some twenty-six Russian military jets operated around Iceland. And in 2011 two Russian Blackjack bombers flew from the Kola Peninsula around Iceland toward North America. More recently, Russia performed an airborne ASW exercise in the GIUK gap just south of Iceland.[4] Norway saw a fivefold increase in Russian air activity, including by bombers and MPAs, over the Norwegian Sea in 2007, in comparison to the previous few years.[5] Some of the Russian sorties were clearly simulated attack runs against Norway's joint command-and-control center in Bodø. The United Kingdom's Royal Air Force, meanwhile, intercepted some twenty-one Russian bomber flights between July 2007 and April 2008.[6] But these early muscle movements of the Northern Fleet in the North Atlantic left little impression on the United States and NATO and were generally deemed as the legitimate maritime activities of a naval power of Russia's size and stature.[7]

U.S. warships seemed to be the priority for Russian prodding and assertiveness in the maritime domain following the Ukraine crisis. In late 2014 the Northern Fleet claimed to have chased a U.S. submarine on a reconnaissance mission out of the Barents Sea, an incident that has been denied by the U.S. Navy. The USS *Donald Cook* was aggressively and closely overflown by Russian fighters in the Black Sea and later experienced the same thing while deployed in the Baltic Sea. During the Baltic Sea deployment, the crew of the *Donald Cook* was ready for what was coming and managed to capture the aggressive and close proximity overflights with handheld cameras and cell phones from the bridge. The dramatic pictures and video of the two Russian Su-24 attack jets roaring by the bridge of the *Donald Cook*, at one point coming as close as thirty feet from the bridge wing and being close enough to the surface of the sea to cause wakes, made international headlines. While close encounters between Russian and U.S. aircraft and warships have been part and parcel of the competition at sea going back to the Cold War, they can have deadly outcomes, as described chapter 5. After the United States and NATO had spent almost three decades operating nearly undisturbed in the global maritime domain, the challenge from Russia caused considerable stir in Washington and

Brussels. But Russia's new level of activity in the maritime domain was not restricted to seemingly intimidating NATO in the North Atlantic and the Baltic Sea. Russian naval power was also brought to bear in Russia's combat operations in Syria, and this effort too would include the Northern Fleet.

On October 15, 2016, Russia's only aircraft carrier, the *Kuznetsov*, slipped its moorings in Kola Bay and headed into the Barents Sea, along with its escort of a *Kirov*-class nuclear cruiser and two *Udaloy*-class ASW cruisers. Three Russian oilers and a tug boat were also part of the *Kuznetsov* group as it made its way into the Norwegian Sea. Belching black smoke that could be seen from miles away, the *Kuznetsov* was on its way to the Mediterranean, but this cruise would be different from her earlier visits to the Mediterranean in 2008, 2011, and 2014. Unlike the previous deployments to the Mediterranean, this was one was intended to be a combat deployment, where the *Kuznetsov* aircraft would come off the carrier to strike targets ashore in Syria. This was part of the Russian campaign to support the Bashar al-Assad regime against the rebels, some of them U.S.-supported, who had sought to overthrow the regime in the wake of the Arab Awakening that had begun in 2011 in Tunisia.

The 2016 *Kuznetsov* deployment raised eyebrows among NATO countries, and a Norwegian P-3 monitored the *Kuznetsov*'s transit of the Norwegian Sea, with the MPA later joined by a Norwegian frigate. As the Russian carrier group entered the North Sea, the monitoring was picked up by two frigates from the British Royal Navy and the Belgian navy. After passing through the English Channel the *Kuznetsov* and her escorts and support ships turned south and sailed past France and the Spanish and Portuguese coasts and entered the western Mediterranean. The *Kuznetsov* group planned to refuel and resupply in a Spanish port while on its way to the eastern Mediterranean, but pressure from NATO and many of its members led the Spanish government to bar the *Kuznetsov* from entering the port. Instead, the carrier and its escorts anchored off North Africa and took on additional fuel from its accompanying oilers, a process that is considerably more difficult and time consuming than refueling at a pier.

The *Kuznetsov* finally arrived off the coast of Syria in late October and began flight operations shortly thereafter. The results were less than what many in Russia had hoped for, and the *Kuznetsov* lost two of its fighter-bombers in crashes as they returned to the carrier after missions over Syria. Later, a complement of the *Kuznetsov*'s Su-33s and MiG-29Ks were transferred ashore to conduct additional strike missions in order to save them from having to conduct carrier recovery operations at the end of their missions.[8] The *Kuznetsov* deployment was the object of substantial media attention, with several European tabloids providing breathless and alarmist headlines and content such as the Russian carrier was "menacing" and carried out "provocative maneuvers" aimed at NATO while on its way to the eastern Mediterranean. Russian media outlets such as Sputnik and Russia Today, often used by the Russian government for disinformation campaigns, also helped fan the flames by providing daily coverage of the feats of the ship and aircrews of the *Kuznetsov* group. While the media coverage at the time may have overdone the immediate threat to NATO nations by the *Kuznetsov*, it did serve as yet another unmistakable signal that Russia was serious about restoring its sea power and about using it as an element of national power both in peacetime and in a crisis. And while the loss of two aircraft and the transfer of aircraft to onshore airstrips point to the Russian navy's continued difficulties with carrier operations, the deployment of the *Kuznetsov* to the Mediterranean for combat operations should not be dismissed as a mere public relations maneuver by Moscow. Carrier operations are arguably the most difficult and complex effort undertaken by any navy, and only a handful of naval powers can be said to fully master the art and science of it. Indeed, America's current superiority in carrier operations rests on a hundred years of diligent work, lessons learned, trial and error, and significant investments. Thus the 2016 combat deployment of the *Kuznetsov* and its escorts, however halting, improvised, and fraught with difficulties, points to the ambition and growing capabilities of the Russian navy in general and the Northern Fleet in particular.

High Kalibr Threat

In addition to the *Kuznetsov* deployment, the Russian navy's submarines would also join the fray in the Mediterranean, again with the Northern Fleet playing a role, even if only a supporting one. In early November 2015 the Russian improved *Kilo*-class submarine *Rostov-na-Donu* finished up test firings of its SS-N-30A Kalibr cruise missiles with the Northern Fleet and made its way from the Barents Sea, around Europe, and toward the Black Sea to join Russia's fleet there and the naval base in Sevastopol in Russia-annexed Crimea. But before entering the Black Sea, the *Rostov-na-Donu* lingered in the Mediterranean, close to the coast of Syria.

The improved *Kilo* submarine is a further development of the Cold War conventional submarine class with the same name, which first came into service in the early 1980s. It is a large boat, with a length of 240 feet and a submerged displacement of around 4,000 tons, making it considerably bigger than, for example, the modern *212*-class boats operated by the German navy. Along with a new propulsion system and upgraded sensors, the main difference between the original *Kilo* class and the improved version is that the latter is also able to fire the Kalibr cruise missile. The improved *Kilo* class has also turned into quite the foreign sales success for the Russian naval industry. At least eighteen improved *Kilo*-class submarines have been sold to China, Algeria, and Vietnam.

The *Rostov-na-Donu* submarine was built in Saint Petersburg on the shores of the Baltic Sea and was delivered to the Russian navy in late December 2014. In August 2015 the Russian navy announced that the *Rostov-na-Donu* had successfully fired Kalibr cruise missiles while submerged at a test range in the High North and was now preparing to join the Black Sea fleet as an operational submarine.[9]

After entering the Mediterranean the *Rostov-na-Donu* took up a position off the coast of Syria and prepared to do what the submarine had practiced in the Barents Sea. On December 8 the submarine launched a volley of Kalibr missiles against targets near Raqqa in Syria, the stronghold of the Islamic State in Iraq and Syria, making it the first time that a Russian submarine had fired cruise missiles in anger.[10] After successfully launching its Kalibr missiles, the *Rostov-na-Donu* passed through the Bosporus while trailed by a Turkish patrol boat, there to monitor the Russian flow of warships from the Black Sea to the Mediterranean and back. The Turks dubbed the regular movement of Russian warships from the Black Sea to the Mediterranean "the Syria express," and the citizens of Istanbul could view the transit and snap photos of the submarine with their cell phones from the shore of the narrow Bosporus that connects the Black Sea to the Mediterranean.[11]

The cruise missile shots from the *Rostov-na-Donu* would be followed by additional volleys fired from both surface ships and other *Kilo*-class submarines in the coming months and years. Later Russia's use of Kalibr cruise missiles in its Syria campaign became more sophisticated, with multiple submarines firing volleys at the same time against a spread of targets.[12] This Russian use of cruise missiles from submarines surprised many in the U.S. and European national security communities. If Russia could launch Kalibr missiles against targets in Syria, Moscow had also shown, by implication, that it could hold big portions of Europe at risk with its long-range submarine-launched cruise missiles.[13] And the *Rostov-na-Donu* Kalibr shot against targets ashore in Syria was not the first time Russia had displayed its growing ability to reach out and touch distant targets with its navy. In October 2015, corvettes in the Caspian Sea fired their own volley of twenty-six Kalibr missiles that traveled over Iran to hit targets in Syria, although some of them fell short and landed in northern Iran. As if the firepower demonstration was not enough of a warning to NATO, the deputy commander of Russia's navy, Vice Admiral Viktor Bursuk, publicly commented at the time that "the range of these missiles allows us to . . . engage targets located quite a long distance away, . . . [this] has come as an unpleasant surprise to countries that are members of the NATO block."[14]

Russia's Kalibr missile is not unlike the U.S.-developed Tomahawk missile, which today is the mainstay weapon for the U.S. Navy for long-range strikes from the sea against land targets. The Kalibr is

an improvement on the Soviet-era Granat cruise missile, first introduced into Soviet service in 1984, which was indeed developed as a
direct response to the Tomahawk system. The original Kalibr missile
had a modest range of roughly 200 miles, and current export versions of the system still have this limited range. Newer land-attack
versions of the Kalibr system, intended for the Russian military, have
far longer ranges, perhaps up to 1,600 miles, although confirmed
figures are hard to come by. Later versions of the missile system can
also achieve supersonic speeds during the end run toward a target.
Originally designed as an antiship missile, the system has evolved
into a family of missiles that includes the aforementioned version for
attacks against land targets, an antisubmarine variant, and a version
capable of carrying a nuclear warhead. All in all, more than ten versions of the Kalibr missile have been developed and produced. While
they are used by the Russian submarine force, there are also versions
for surface warships and aircraft, as well as in coastal batteries. The
latter version has also been packed into standard shipping containers, which means it could be rapidly moved by rail, truck, or cargo
ships in a tough-to-spot way and later emerge as part of a Russian
defensive network in a zone of crisis.[15]

The Kalibr missile is nearly 26 feet in length and carries a 1,100-
pound conventional warhead, nearly double the weight of the warhead on the Tomahawk. After launch from a submarine's torpedo
tube or from a vertical launch system on board a surface ship or
submarine, the Kalibr flies toward the target at an altitude of around
two hundred feet. It cruises along at a speed that is a little faster than
a modern jetliner, at about 615 miles per hour. As the Kalibr nears
its targets it enters into attack mode and sprints the last distance at
about Mach 3, or 2,300 miles per hour.

The importance of the Kalibr system in modern Russian naval
warfighting should not be underestimated. The strikes against Syria
using Kalibr missiles from a conventional *Kilo*-class submarine in
the Mediterranean and corvettes in the Caspian Sea show that even
smaller Russian warships now pack a real punch that can reach
far across the maritime domain and onto land.[16] The use of cruise

missiles from submarines and surface warships also puts Russia in an exclusive club, which until recently included only the United States and the United Kingdom. Even in the 1990s the use of cruise missiles was considered not only too sophisticated but prohibitively expensive for powers other than the very richest major powers. But the proliferation of technology, in particular accurate navigation systems and ever cheaper computing power, means that cruise missiles are increasingly available at a reasonable cost to emerging and revanchist powers, including Russia. Cruise missiles have been key elements in U.S. operations since the early 1990s, both as a rapid response to an emerging crisis and to deter unwanted actions by an adversary. Cruise missiles fired from American and British submarines have also played key roles in the opening phases of major military operations, such as the Gulf War, the Kosovo campaign, the Afghanistan war, and the invasion of Iraq in 2003.

A cruise missile is also particularly difficult for a defender to contend with. First, unlike manned bombers or fighter-bombers it is not difficult for cruise missiles to maintain low altitude flight, which means that it is more difficult for radars to pick out the incoming cruise missile from other ground clutter. Furthermore, modern cruise missiles can be preprogrammed to fly routes that evade radar coverage. Cruise missile attacks from several platforms firing from several directions can also be coordinated, which could overwhelm air defense systems. Second, cruise missiles, such as the Kalibr, are relatively small, which makes for a small radar cross section, far smaller than a manned aircraft. Finally, the engine on the Kalibr, and other cruise missiles, is relatively small and placed in the back, which means that sensors would have a hard time picking up heat signatures at any real distance.[17]

Cruise missiles can be delivered to a target from a range of platforms: aircraft, surface ship, or ground launcher. But launching a cruise missile from a submarine carries special advantages. An aircraft or surface ship will often have to release its cruise missiles at the outer edge of its effective range, or risk a counterattack by the defender. This will increase the flight time of the cruise missile and increase the chances to mount an effective defense against the attack. Also, an aircraft or surface ship is still likely to be detected and tracked by the defender, even if it remains outside the range of air defenses or coastal-based antiship missile defenses, which gives the defender advance warning of an impending cruise missile strike. Not so with a submarine. It can carry a cruise missile closer to its intended target while remaining undetected, which means that the cruise missile's flight time can be greatly reduced, which in turn reduces the chances of mounting an effective defense against it. A submarine can therefore also conduct cruise missile attacks in a sudden way that gives the defender little notice.[18] This is not to say that the submarine has all the advantages when it comes to using cruise missiles. First, the launch of cruise missiles from a submarine will nearly always give away the location of the submarine. In the case of the Kalibr system, the launching submarine must extend a mast above the surface of the sea to receive targeting data ahead of the launch, which provides a defender at least a fleeting opportunity to detect the launch platform.[19] The launch itself will give ASW forces a strong cue for further search for and prosecution of the target. This means that the submarine will have to quickly leave the launch area to escape being hunted down. Second, the ability of submarines to carry cruise missiles is limited, especially in the case of the smaller conventional submarines.

The use of cruise missiles as antiship weapons from submerged submarines is still, however, a relatively complicated affair. The submarine's sensor range remains relatively short, and it certainly falls far short of the horizon, which negates a significant advantage of cruise missiles: the ability to attack surface ships over the horizon. Therefore, for targeting its antiship cruise missiles against faraway ships, the submarine still needs input from other sensors borne by surface ships, aircraft, or at sites ashore. This sensor data needs to be transmitted to the submarine and fed to the cruise missile before it is fired. The missile's own tracking system can take over once the cruise missile nears the target and can detect the intended ship with its own sensor suite. The effective use of antiship cruise missiles therefore needs a robust network of sensors, command-and-control, and launch platforms, in this case submarines. Striking land targets with cruise missiles from a submarine is, in comparison to attacking ships at sea, a relatively easier endeavor. Major infrastructure, such as ports, naval and air bases, and command-and-control centers, are immobile targets with easily obtainable locations.[20] A cruise missile can therefore be preloaded with the target location and guided to the target by GLONASS, Russia's own GPS-like satellite navigation system, and with on-board terrain recognition systems. Both of these methods are used by the Kalibr.

A range of Russia's nuclear and conventional submarines can carry the Kalibr cruise missiles, including the aforementioned improved *Kilo*-class submarine, which is in service with both the Northern and Black Sea Fleets. The Kalibr is also carried by the *Oscar-II* SSGN class, a submarine first introduced in the late 1970s but which has seen upgrades since that time in order to, among other things, accommodate up to seventy-two Kalibr missiles.[21] Currently six *Oscar-II*–class submarines are in service with the Russian navy, of which four belong to the Northern Fleet. The Kalibr is also the mainstay weapon for Russia's new *Severodvinsk* class, a multipurpose SSN that can carry up to forty Kalibrs in its vertical launch system. Laid down in 1993, the first of these boats was taken into service in the Northern Fleet in 2013, with the second boat, the *Kazan*, expected to enter active service in 2018.[22] The christening of the *Kazan* SSN, which itself includes improvements in comparison to the first-in-class boat *Severodvinsk*, came with a sleekly produced thirty-minute online video that compared the new submarine with both the U.S. *Seawolf* and *Virginia* classes, along with extended shots of *Severodvinsk*-class submarines firing both land-attack and antiship cruise missiles. Along with Kalibrs, the *Severodvinsk* class can carry other antiship and land-attack missiles, such as the P-800 Onyx, and the hypersonic cruise missile Tsirkon, which is currently under development in Russia.[23] Russia is also at work developing a new SSN, for now dubbed Project Husky, which would rely on simpler solutions and proven technology to make it more affordable than the *Severodvinsk* class. The Husky submarines, while not expected to be as capable as the *Severodvinsk*, will likely be dedicated to defending the Russian SSBNs, a duty for Russian attack submarines first seen during the Cold War. This approach would free up the more advanced *Severodvinsk* boats for strike operations in the North Atlantic and elsewhere.[24]

The combination of submarines and cruise missiles, whether land attack or antiship, also has real implications for ASW. Traditional ASW occurs at relatively close range as the submarine must approach its intended target at slow speeds to bring its torpedoes to bear. This means that the engagement zone for ASW forces is relatively small once the battle has been joined, and short-range helicopters and surface ships can be effective using antisubmarine torpedoes. But the use of cruise missiles means that the submarine can stand off from its intended target, whether it is a warship, a merchantman, or a target on land. The ASW search-and-engagement area thus widens considerably, and short-range and slow-moving ASW platforms such as helicopters and frigates will have a challenge in detecting and engaging the submarine.[25]

The Russian effort of pairing relatively modest platforms with long-range cruise missiles, and the integration of cruise missiles into the coming classes of Russian submarines, have increasingly caught the eye of Western observers and militaries. A 2017 Defense Intelligence Agency report on Russia's military power stated that "the proliferation of this capability within the new Russian Navy is profoundly changing its ability to deter, threaten, or destroy adversary targets."[26] The Center for Strategic and Budgetary Assessments, a respected and influential Washington think tank that has performed a number of deep studies on the future of warfare for various offices at the Pentagon, reported in 2011 that due to the proliferation of cruise missiles and other precision munitions and their distribution to American adversaries in the future, "large or massed ground forces, major ports, and bases are likely to become highly vulnerable to enemy . . . missiles."[27]

This approach would also be especially attractive to Russia considering the limited number of ports and airfields that are currently available to accept, process, and send American reinforcements on their way toward eastern Europe in a crisis.[28] According to United States European Command, there are currently only seven ports in northwestern Europe that are fully capable of accepting heavy U.S. reinforcements coming across the sea, including four in Germany, one in the Netherlands, one in Belgium, and one in the United Kingdom. There are six key air nodes in northwestern Europe, with half of them within the reach of cruise missiles launched from the Norwegian Sea or the North Sea. But while many of the air nodes cannot be reached with long-range weapons from the sea, much of the heavy equipment and vehicles needed to reinforce Europe in a crisis would not be able to be flown in. They would have to be sailed across the Atlantic or already be forward-staged on the continent.

This approach of using submarines and surface ships for short-notice long-range strikes against land targets to intimidate an adversary or disrupt reinforcements also broadly aligns with Russia's understanding of war in the modern age. Senior Russian military leaders have publicly noted the dynamic of a rapidly emerging crisis than can quickly develop into a sharp war. Russian leaders have also noted the frequent use of accurate long-range strikes by the United States and Britain, often from submarines, to signal an adversary, deter an enemy's actions, or open a conflict with a decisive blow to an opponent's key capabilities, such as logistics, command-and-control, or air defense systems. The Russian military doctrine therefore puts a strong emphasis on long-range strike capabilities that can pre-empt the actions of an adversary.[29] Indeed, the Russian sense that the combination of long-range weapons, powerful sensors, and computer-based command and control is a potential game changer goes back to the late Cold War and was seemingly confirmed in Moscow's mind with the rapid American victory using exactly this strategy during Operation Desert Storm in 1991.[30] And the notion of attacking ports and airfields in a bid to halt NATO's reinforcement operations, instead of waging an antishipping campaign, also has Soviet roots, even though the original concept was never realized during the Cold War due to technological limitations.[31]

The use of cruise missiles to threaten NATO's reinforcement efforts across the Atlantic by destroying the ports needed to accept and process them also fits squarely into Russia's broader anti-access, area-denial (A2/AD) strategy it has been developing over the past decade to hold NATO at a distance from regions of military interest to Russia. Based on long-range ballistic, air defense, and antiship missiles stationed on key pieces of terrain to frustrate America's and NATO's ability to quickly reach a region of interest (or threatened ally), these Russian A2/AD networks have emerged in the Kaliningrad enclave in the Baltic and on Crimea in the Black Sea. These A2/AD networks can be further extended by using other assets to contest the airspace and the sea, for example fighter jets and in particular submarines. Russia also sped advanced air defense and shore-based antiship missiles to Syria in 2016 to help deter the West from considering any sort of military intervention in Syria to help topple the Assad regime during the civil war there.[32]

A strategy of denying access to a crisis area is a logical and reasonable way forward for powers such as Russia. It has closely watched more than two decades of U.S. military operations around the world where the United States and its allies have been able to build up combat power for decisive engagements almost undisturbed by the opponent, be it Serbia, Iraq, or Libya. The A2/AD debate has been ongoing in American circles for years and has in particular focused on the Asia-Pacific region and Chinese military developments. It has also elicited U.S. military concepts to defeat it, such as the Air-Sea Battle framework, but the notion of area denial came roaring over to Europe when it was clear by mid-2014 that Russia was seeking to break the European security order. Since then A2/AD has been on the lips of NATO officials and senior military leaders in Brussels as one of the major challenges facing the alliance from a resurgent Russia. And this has included senior U.S. military commanders from U.S. European Command and the services. In early 2016 the commander of U.S. European Command, Gen. Philip Breedlove, told assembled journalists at the Pentagon that rapid reinforcement of certain NATO nations "cannot be taken for granted anymore" due to Russia's A2/AD networks.[33]

While Russia's ambitions for its developing A2/AD networks in the Baltic Sea and the Black Sea seem relatively contained to those two regions, that is not the case for the Kola Peninsula and the North Atlantic. Here, the A2/AD concept harks back to the Cold War days of bastion defense, when the Soviet Union sought safety for its SSBNs behind screens of air defense systems, fighter jets, attack submarines, and ASW frigates. Many indications point to Russia's seeking to expand its ability to deny the sea and air domain to the United States and NATO well into the North Atlantic, and perhaps all the way to the GIUK gap. And this extended bastion system is no theoretical concept. It has been exercised in the real world. In mid-March 2015, Russia launched an unannounced snap exercise that developed into one of the most far-ranging ever put on by the Russian military. All in all, it included some 80,000 personnel with exercise activities from the Baltic Sea region to the Black Sea and the Pacific. The exercise clearly simulated a conflict with NATO, and the first stage of the exercise was concentrated in and around the Kola Peninsula, where the Northern Fleet was deployed to conduct ASW and surface warfare exercises in the Barents Sea. At the same time, Russian strategic bombers performed strikes against simulated targets on the Kola Peninsula, while air defense units simulated defense against incoming air threats. Marine infantry assigned to the Northern Fleet were also mobilized and moved into the Arctic. During the second and third phase of the exercise, Russian ground, air, and naval units in the Baltic Sea region and around the Black Sea joined in the exercise to conduct live-fire drills in their area of responsibility.[34] General interest media outlets called the Russian focus on the Kola Peninsula during the exercise "utterly baffling," since the territory there is well away from the heart of Europe and with a low NATO presence in the surrounding region.[35] It is not so baffling if one transplants the events in and around the Kola Peninsula and the Barents Sea in mid-March 2015 into the Atlantic and the Norwegian Sea at some future date. The first phase of the exercise was clearly a rehearsal on how to deploy the extended bastion in order to deter and hinder NATO reinforcements coming into Europe across the Atlantic, strike land-based targets that would support NATO operations in the Atlantic (e.g., as ports, command centers, and intelligence nodes), and provide strategic cover for Russian ground, air, and maritime operations elsewhere. The same type of operation in the North Atlantic was rehearsed again during Russia's major and recurring exercise Zapad in September 2017.[36]

The introduction and dissemination of cruise missiles across the Russian navy will make the emerging fourth battle of the Atlantic different from the previous three. The resurgent Russian submarine force, with its emphasis on the Northern Fleet, is numerically too small to be able to mount an effective antishipping campaign in the vastness of the North Atlantic south of the GIUK gap. An effective antishipping campaign in the North Atlantic would also have to rely on additional cruise missile attacks from the air, as during the late Cold War—something that today's Russia is unable to mount given the aging Russian bomber fleet and the lack of a replacement bomber in sufficient numbers.[37] But an antishipping campaign in the North Atlantic conducted by both submarines and naval aviation bombers is no longer necessary if the fewer but more lethal Russian submarines can put their Kalibr missiles to work against the relatively limited number of ports and airports that U.S. and NATO reinforcements will have to rely on to get reinforcements onto the continent and on their way to eastern Europe.[38] Cruise missile attacks could also be used to strike at NATO's command-and-control centers in northern Europe, or at key ISR sites, such as the Vardo radar site in northern Norway.[39] Furthermore, even the deployment of cruise-missile-armed submarines into the North Atlantic could serve as a deterrent against NATO action, as they could be launched on short notice against population centers or other vital areas. And the use of submarines, rather than surface ships or aircraft, would add an element of uncertainty that could be difficult for political decision makers to deal with, as it would be tricky to truly know the urgency or scope of the threat and whether NATO navies have been able to detect and track them all before a strike.[40] With submarines armed with cruise missiles, the Northern Fleet can remain well north of the

GIUK gap, and thereby avoid the maritime choke point and NATO and U.S. surveillance there, and be closer to its home bases and still be effective in the pursuit of disrupting NATO reinforcement efforts or deterring allied action.[41]

As a final note on the Kalibr system, it is important to point out that the rate of production of the Kalibr system for the Russian navy has remained relatively low, and at the current rate it will be years before Russia reaches the level of on-hand inventory assumed to be needed for a major conflict, which is roughly five thousand weapons.[42] The production of Kalibr missiles could of course be accelerated, and new production facilities could be opened to speed the building of the inventory further. This is one area that bears watching in the coming years, as one indication of Russia's intent and preparedness for major combat operations at sea.

The Vulnerable Submarine Cables

The return of the Russian navy to the open seas has also meant a greater Russian focus on maritime intelligence gathering. Here Russia has taken a special interest in the submarine cables in the North Atlantic that connect North America and Europe. By some measures Russian subsurface activity around known submarine cable routes across the North Atlantic, and also in the North Sea, now rivals the levels seen during the Cold War.[43] The Russian interest in submarine cables across the Atlantic is thus not entirely new. The presence of Soviet intelligence ships around cable stretches had been noted as early as 1960, and a Soviet intelligence-gathering ship was boarded and searched by a U.S. destroyer off Newfoundland in 1959 after a number of AT&T cables were disrupted, although the search proved inconclusive.[44]

The first submarine cables were drawn between Europe and North America in the late nineteenth century. At first, the information that could pass through them was extremely limited. In the early twentieth century the transatlantic submarine cable system was able to pass, through telegraphy, the modern equivalent of roughly eight megabytes of data annually, or far less than the size of the average thirty-second

YouTube video.[45] The early cables were brittle and would break often, requiring frustrating repair work at sea. Today, the cables are far more hardened, but they still break from time to time. The global submarine cable network has also grown dramatically since the early twentieth century and now provides communications linkages across all the major oceans and beyond. The introduction of satellite-based communications in the 1960s caused a brief decline in the use of submarine cables for global connectivity, but the commercialization of fiber optics in the 1980s, which increased data capacity by an order of magnitude, put much of global communications back underwater and inside submarine cables.[46] Indeed, today the vaunted communications satellites in orbit around the earth handle only around 5 percent of the global communication flow, while the other 95 percent goes through undersea communications cables. In addition, data transmitted by satellite only travels at about a fifth of the speed achieved by submarine cables.[47] The private sector is far from the only user of the fast and high-bandwidth communications provided by submarine cables. Due to the global nature of American power, U.S. military and allied operations also rely on submarine cables for the fast transfer of huge amounts of data. This reliance has increased further with the growth in the use of unmanned reconnaissance and strike systems and with the near-instant feed of data preferred over the slower communications offered by satellites. The slower speed of the latter causes a lag between the input of an operator in, say, Nevada and an unmanned system high above the Middle East. Indeed, a single Global Hawk drone, the U.S. military's long-range and high-altitude unmanned reconnaissance vehicle, requires a whopping five hundred megabits per second of bandwidth for its operations. And a sizeable portion of U.S. drone data is funneled through Ramstein Air Base in Germany and then via submarine cable across the North Atlantic to various installations and commands in the United States.[48]

All in all, the world's oceans contain nearly a million miles of undersea cables, with the North Atlantic home of one of the densest cable networks in the world, which connects North America and Europe at various points. The connections across the Atlantic are especially dense between the U.S. northeast and points in the United Kingdom and France, with the largest capacity cables anywhere in the world strung between London and New York.[49] But the network also includes connections between mainland Europe and Iceland and Greenland, along with cables between northern Norway and Svalbard, which means that the inhabitants of even this remotest of outposts on the edge of the North Atlantic can enjoy broadband Internet access and excellent cell phone service.

The Internet may be distributed and hard to completely take down (a downed node in one place can easily be rerouted around), but that is not so at sea, which is effectively an infrastructure bottleneck. Only a little more than two hundred systems funnel the world's Internet traffic across the oceans. And more submarine cables are coming, including in the North Atlantic. To support its budding business in cloud services, Microsoft and Facebook announced in 2016 that the two companies would partner on a new 3,700-mile submarine cable from Virginia Beach to Bilbao in Spain, which would carry the immense amounts of data produced and demanded by Microsoft and Facebook customers and users.

The planning and laying of the cable routes across the Atlantic and the landing spots used in Europe and North America during the Cold War were to some degree informed by national security considerations in order to improve redundancy and resiliency of communications during wartime. Today, most new cables are laid and maintained by international corporations, with the routes and landing sites informed by efficiency and financial bottom lines, which potentially leave them more vulnerable to tampering or invasive intelligence gathering.[50]

Disrupting or intercepting enemy communications has always been a key feature of warfare, and so it is with submarine cables as well. A U.S. Marine raiding party cut the cable that connected Cuba to Spain during the Spanish-American war in 1898.[51] Shortly after declaring war against Germany in 1914, Britain dispatched cable ships to cut Germany's submarine telegraph cables, leaving Germany with only one active cable, which the British tapped into. The operations to cut the cables were crude by today's standards, involving hauling the cables on board with grapples, and then simply chopping them off with hatchets and dumping the ends back overboard.[52] The British attacks on the German communications links sparked a "cable war" between the Royal and German navies, a lopsided contest as Germany lacked most of the tools and the types of ships to reach and cut the subsurface cables relied on by the British during World War I.[53] A German U-boat did, however, manage to sever the connection that provided telegraph service between the United Kingdom and Portugal, the Azores, and Gibraltar.[54] Likewise, Nazi Germany's submarine cable connection across the North Atlantic to North America, routed through the Azores, was severed shortly after the beginning of World War II. But to completely sever the communications connection provided by submarine cables between Europe and the United States today would be hard. The cable system is far denser than the four cables that Germany had at its disposal during World War I, and communications can be rerouted around broken submarine cables and on to intact ones. Satellites can play a role in taking some of the pressure off the submarine cables, but it is expensive and would lead to lag times in the data transfer. Cutting a cable would be disruptive nonetheless, as was the case in 2007 when Vietnamese salvagers inadvertently pulled up two submarine cables off the coast of Vietnam, which slowed access to the Internet in that country for months. U.S. military operations were impacted in December 2008 when an anchor dragging across the sea floor in the Mediterranean severely bent three submarine cables connecting Europe with Egypt. This disruption slowed U.S. unmanned aerial vehicle operations out of Balad air base in Iraq from hundreds of sorties a day to only a handful until the cables could be fully repaired.[55] Submarine cables also continue to be attractive targets because of the rich intelligence from both corporations and governments, especially in an age when most institutional operations rely on access to cyberspace. Indeed, tapping into Soviet communications cables in the Barents and Bering Sea proved to be one of the largest U.S. intelligence-gathering coups of the Cold War.[56]

Russia maintains a fleet of auxiliary submarines for these types of undersea missions, with the majority of them homeported with the Northern Fleet. They can dive deeper than most submarines and reach infrastructure on the ocean floor, but the gain in deep diving has meant that the special mission submarines, such as the AS-12 *Losharik*, have had to give up endurance and range; they will have to be carried to the mission area by motherships, such as converted SSBNs. Russia has converted at least two SSBNs for this purpose, including the *Delta III*–class *Orenburg*, which entered the Northern Fleet in 1981 and began its conversion in 1994 by removing its missile compartment and replacing it with a larger holding compartment.[57] The *Orenburg* and the *Losharik* worked together in 2012 during a Russian Arctic expedition to perform ocean-floor surveying that would help bolster Russian claims to the Arctic sea floor, with the *Losharik* reportedly working at depths in excess of eight thousand feet for long stretches of time.[58] In 2014 the *Orenburg* was, by happenstance, sighted again in the Arctic by Norwegian researchers doing scientific work on an ice floe.[59] And one of Russia's mothership submarines was tracked by the French navy in the Bay of Biscay in early 2016. Recent reports also suggest that Russia is converting the unfinished *Oscar-II*–class submarine the *Belgorad* into a special-mission vehicle, able to carry unmanned vehicles and lockout chambers for divers, and will include large compartments to carry bulky equipment to the ocean floor.[60] Russia also seems to be developing its auxiliary submarine force further. The new *Sarov B-90*–class submarine was developed and built by the Sevmash shipyard on the Kola Peninsula. It has a curious propulsion system, consisting of diesel-electric machinery along with a nuclear reactor that is not mechanically connected to the propulsion system. The *Sarov* is thought to be a testbed for new technologies, but it is also reported to be able to deploy unmanned underwater systems and mines.[61]

The West's reliance on information networks, coupled with Russia's growing fleet of specialized submarines, means that the submarine cables across the Atlantic represent a key vulnerability for NATO and the United States in the new contest with Russia. This vulnerability is far larger today than during the preceding three battles for the Atlantic.

Russia's Naval Modernization

Russia's increasing level of naval activity is powered by its ambitious navy modernization program, which includes not only a robust shipbuilding plan but a transition to an all-volunteer force of sailors and expansion of supporting infrastructure. This effort, with a particular focus on the Northern Fleet and the Black Sea Fleet and the undersea domain, is beginning to pay dividends. Progress, however, has been by no means in a straight line, and the advancements have been marred by corruption, technical difficulties, and incidents that have resulted in public embarrassment both ashore and at sea. In late 2018, for example, the carrier *Kuznetsov*, back from her deployment in the Mediterranean, was damaged during deep maintenance when its dry dock flooded and a crane smashed into her. Still, the investments have paid off to a degree that many Western analysts have failed to foresee. Indeed, mocking the apparent shortcomings of the Russian navy has apparently become a pastime among some Western Russia watchers.[62] And while the shortcomings are real, they should not be overstated to the point that they obscure the very real progress being made.

In its modernization efforts, the Russian navy, along with the other services of the Russian armed forces, faces a set of real challenges, including the need to recruit and retain talented and skilled personnel from a shrinking and in part unhealthy population base, the urgent need to modernize and capitalize the Russian naval industrial base, securing financial resources over the long term to sustain the modernization, and attracting foreign know-how and participation in key naval projects.[63] These would be real and significant challenges for any nation seeking to modernize its navy. It should be remembered, however, that even the United States and its Western allies have at times struggled with recruitment, financial sustainability, and getting major procurement projects completed on time and on budget. Indeed, the

U.S. Navy is facing a looming maintenance crisis as it continues a high pace of operations after more than a decade of wars in the broader Middle East. This is not to equate the challenges of the Russian navy and those of the United States and its allies, but only to say that the Russian struggle to develop and use its maritime power is far from unique.

Along with the already described improved *Kilo* SSKs and the *Severodvinsk* SSNs, the Russian naval industry is also hard at work bringing online submarines with air-independent propulsion (AIP) technology, with the first class of AIP boats being the *Saint Petersburg* class. AIP-technology submarines are by themselves no longer a new phenomenon. Several submarine-operating nations have developed AIP technologies, either based on fuel cells or Stirling engines. Up until recently, however, the nations with AIP submarines have been either NATO countries and U.S. allies or at least close partners of the alliance and the United States, such as Sweden. Now both Russia and China seem to have AIP submarines within their reach after years of development and trial and error. Indeed, AIP submarines in the hands of an adversary has been a real worry for the U.S. Navy for many years, with the technology being, for example, one of the fields that is closely followed by the annual Report on the Military Power of the People's Republic of China. AIP is a real leap forward in terms of the capabilities of conventional submarines. With AIP propulsion, conventional submarines are no longer forced to come to snorkel depth to recharge their batteries, making the boats especially vulnerable to detection by surface and airborne ASW units. Subsurface endurance has increased significantly with AIP, from a day or two to closer to a month. This makes ASW more taxing and a conventional boat a more formidable opponent.

When comparing the advances in the subsurface domain to those on the surface, Russian shipbuilding has not produced much in the way of major surface warships. Russia's last major surface combatant was the nuclear-powered battle cruiser *Peter the Great*, which was sent to Severomorsk in 1998 to serve as the flagship of the Northern Fleet. Construction of the *Peter the Great* began in 1986 and took nearly twelve years to complete, which says more about problems ashore at the yard than about the complexity of the actual ship. In 2000 the cruiser was on site during the *Kursk* catastrophe, as it was intended to be the target of the *Kursk*'s simulated torpedo attack. The other major surface ship produced in recent times by Russia is the *Kuznetsov* aircraft carrier, which entered service in 1995.[64] By comparison, the U.S. Navy has commissioned four aircraft carriers since 1995. Russia's shipbuilding plan is instead focused on smaller ships rather than larger destroyers or aircraft carriers. But to think this means an exclusive Russian focus on coastal operations close to home would be a mistake. With modern missile technology, such as the Kalibr missile, even smaller ships can pack a wallop and reach far into the maritime domain and onto relatively distant shores. Indeed, during Russia's aforementioned Syria operations, the navy fired Kalibr missiles from the Caspian Sea that traveled across Iran and Iraq to hit targets in Syria, a distance of nearly 1,200 miles.[65]

Along with boosting its submarine fleet, Russia has also sought to grow its amphibious capabilities, including by using foreign suppliers.[66] In 2009 Russia approached the French shipyard DCNS to discuss the possibilities of France, a NATO member, providing two ships of its *Mistral* class to the Russian navy, with one of the *Mistral* ships intended for Russia's Northern Fleet.[67] The *Mistral* class was originally designed for the French navy, which operates two of the ships, and they are capable of carrying some 70 vehicles, 450 troops, and an assortment of helicopters. The deal with Russia would include cooperation around the command-and-control systems on board the ship, along with a transfer design and construction know-how to Russian shipyards. In this pre-Ukraine crisis period concerns were raised only by NATO's easternmost allies, while the deal was primarily seen through a commercial lens in Paris. Indeed, at the time the Russian outreach to a French shipyard was taken as another sign of Russia's maritime weakness. The move was interpreted to mean that the Russian naval industry now needed to import know-how and expertise to complete projects of any meaningful size. But there is a long history in Russia of international outreach to gain access to expertise and practical skills in the maritime sector. Peter the Great certainly did just this in the eighteenth century, and in the 1930s and 1940s the Soviet Union relied on designs and expertise from Italy, France, and Germany.[68] Norwegian and Japanese technology, perhaps illicitly exported, was used to produce more sophisticated propellers for Soviet submarines in the late 1980s, which contributed to the further quieting of Soviet submarines.[69] The *Mistral* deal was finally scuttled in 2015 due to Russia's aggression in Ukraine, with France returning some of the funds Russia had already spent on the class of ships. The two completed *Mistral*-class ships were later sold to Egypt.

Russia's naval resurgence is not only focused on additional submarines, warships, increased exercises, and extended operations. It is also supported by the modernization and new construction of facilities ashore. On the Kola Peninsula the Russian government is constructing new bunkers and auxiliary facilities to store nuclear weapons close to Severomorsk naval base, along with transport infrastructure to speed the delivery of the weapons from storage to the waiting SSBNs.[70] The Northern Fleet is also supported by a new administrative and logistics command introduced in 2012 called the Arctic Center for Material and Technical Support, staffed by some 15,000 personnel.[71] Sevmash, the major submarine producer in Severodvinsk on the White Sea, has also stopped its diversification into civilian shipbuilding and gone back to focusing exclusively on building submarines and warships for the Russian navy and foreign clients.[72]

Some doubt Russia's ability to maintain its pace of military modernization and tempo of exercises and operations. It is true that many of Russia's fundamentals are not pointing in the right direction. Russia is not among the top innovative countries in the world, in spite of active encouragement from the government in Moscow and subsidies for the creation of a Russian-style Silicon Valley. The population is also aging and, in comparison to other industrialized nations, in appalling shape and health, with male life expectancy in Russia hovering around sixty-four years, which places Russia between Senegal and Gabon in a global comparison.[73] Russia is also heavily dependent on its energy sector for revenue. Energy accounts for nearly 70 percent of Russia's export income, a lucrative endeavor when energy prices are high, but a dangerous dependency when prices go lower.[74] The sanctions applied to Russia after the Ukraine crisis have also taken their toll on the Russian economy, leaving many important Russian companies unable to transact business in the West. Still, the Russian economy may be more resilient than many expect. Beyond oil and gas, Russia also has vast pools of other resources, including timber, gold, coal, and nickel. Russia's defense industry also does brisk business around the world, particularly in aviation, armor, and air defense systems. Russia is the world's second largest arms exporter, behind the United States, with a roughly 20 percent share on the world markets and with $50 billion of sales on the books.[75] Finally, Russia also holds a close to $400 billion foreign exchange reserve.[76] Thus, in spite of the sanctions, the reliance on the energy industry, and the lack of innovation, Russia should be expected to continue its military modernization in the coming years, even if at a slower rate.

Russia's Maritime Strategy

Russia's return to the sea in general and the North Atlantic in particular is also informed by a long-term maritime strategy, which has evolved over time as the NATO-Russia relationship has become more strained and Russia more revanchist in posture. Russia unveiled a new "Sea Doctrine of the Russian Federation" in 2001, which outlined Russia's maritime interests and how the government was to proceed to safeguard and develop them further. In tone and focus it was primarily a civilian-oriented document, with a particular emphasis on advancing Russian merchant shipping; protecting, developing, and extracting maritime resources; and advancing the freedom of the seas. Tom Fedyszyn of the U.S. Naval War College, a noted expert on the Russian navy, said about the 2001 document, "It was almost as if there wasn't a Navy admiral present at the drafting. It was like the Secretary of Commerce, the Secretary of Tourism, and the Secretary of Energy sat down and, at the last minute, invited an admiral to come in and write a couple of words. It clearly was a maritime doctrine, not a naval doctrine." Indeed, this first version of a new Russian maritime doctrine would have been cheered by those who thought that Russia may be moving toward embracing the norms and values of the West. The doctrine outlined threats and challenges, such as piracy and interruptions to international sea-lanes, which would have seemed familiar in any Western maritime strategy document at the time, as outlined previously in this book.[77] In 2015 the doctrine was updated, however, and it took on an entirely new character, transitioning from a maritime doctrine to an essentially naval one. The 2015 version included language that established that Russia needed capabilities to stop non-Russian naval vessels from operating in the Arctic during a crisis, along with a need for Russian access to the Atlantic and the ability to stop NATO forces from operating in that maritime domain. The new revised doctrine was also unveiled with a highly choreographed TV broadcast, where Vladimir Putin was surrounded by his defense minister, along with several senior admirals and the Russian chief of defense. The nondefense ministries with a role in maritime affairs, which had been so central to the development of the original 2001 maritime doctrine, were not invited to attend the televised ceremony.[78]

With Russia's new levels of activity at sea in the context of the new security environment, it is important to remember Russia's geostrategic position as it relates to its maritime domain. Russia is traditionally considered a continental power, sitting as it does at the center of the "Eurasian Heartland," a term coined by turn-of-the-century strategist Halford MacKinder. But Russia has also always sought to be a major sea power and recognizes the importance of the oceans in giving Russia global reach. After all, Russia's aggregate coastline is one of the longest in the world, comparable to those of Canada and the United States.[79] Since developing out of the Grand Duchy of Moscow, Russia has extended its reach to the Baltic Sea, the White Sea, and the Black Sea.[80] Tsar Peter the Great founded Saint Petersburg, today one of Russia most important cities, to serve as an anchor point in the Baltic Sea region and by extension the conduit for Russia's interaction with Europe across the sea.

Today, Russia's geostrategic situation has changed considerably since the end of the Cold War, but its ambition to be a major maritime power has not. Geographic conditions, however, funnels that maritime ambition toward the north. Post-Soviet Russia has lost much of its littoral territory around the Baltic Sea, with Poland and the Baltic states now fully integrated NATO members. What remains is the outlet at Saint Petersburg through the Gulf of Finland and the Kaliningrad enclave in the southeastern corner of the Baltic Sea. The Baltic Sea no longer provides Russia with assured access to the broader maritime domain, given that the other Baltic littoral states are either members of NATO or close partners of the alliance. In a war, the Russian Baltic Sea Fleet would find it impossible to exit the Baltic Sea, but it could, on the other hand, certainly cause NATO and the United States significant frustration if they attempted to reach the Baltic States across the sea.

The Black Sea offers little more than the Baltic Sea. Romania and Bulgaria are today NATO members, and the outlet from the Black Sea to the Mediterranean is controlled by Turkey, another NATO member. Russia's navy certainly operates in and out of the Black Sea in peacetime, but in a crisis the Black Sea Fleet would be bottled up and at the mercy of long-range NATO and U.S. air attacks. Russia's maritime possibilities have indeed improved in the Black Sea; with the annexation of Crimea, Russia also assured itself long-term access to the naval base at Sevastopol, which previously was under a long-term lease from Ukraine and which could potentially be lost forever if Ukraine entered NATO as a full member.[81] Access for Russia from the Black Sea to the broader global maritime domain, however, remains relatively constrained. The Pacific also does not offer much in the way of a maritime outlet for Russia, as Arctic ice coverage restricts the approaches and requires relatively close navigation to Japan or Alaska. On Russia's Pacific coast, Moscow must also contend with an increasingly powerful China that is growing and showing its naval muscles. Russia and China have held joint naval drills, both in the Pacific as well as in the Mediterranean and in the Baltic Sea, but it is a wary relationship indeed.[82] Russia also gave up one of its few extraterritorial maritime toeholds in the Pacific when it ended its lease on access to Cam Rahn Bay in Vietnam in 2002. No plans seem to exist to regain access to that base.

Russia's new geopolitical position therefore leaves the European Arctic, the Barents Sea, and the North Atlantic as the avenue for Moscow to express its growing naval and maritime ambitions. Put in a different way, the European Arctic, particularly the Kola Peninsula, and the approaches to the Atlantic across the Barents and Norwegian Seas is the path of least resistance for a Russia that must contend with rising powers to its east and lost territory and influence to its west.[83] Indeed, the opening up of the Arctic due to climate change increases the attraction of the northern maritime emphasis in current Russian strategic thinking. Thus the emphasis on the North Atlantic and Northern Fleet in terms of modernization and assigned units makes eminent strategic sense for a power that knows it is in an asymmetric contest and can ill afford to squander its resources in areas and regions where they will likely not pay dividends.

CHAPTER 11

The Fourth Battle of the Atlantic

The Admirals' Warning

In June 2016 the prestigious naval affairs magazine *Proceedings*, published by the United States Naval Institute, which is influential in naval circles both in the United States and around the world, ran an article by the then-commander of the U.S. Sixth Fleet, Vice Adm. James Foggo. The article, titled "The Fourth Battle of the Atlantic," played on the memory of the previous three struggles for control of the broader Atlantic and its importance to victory in Europe. It rang the alarm bell on Russia's increasing naval activity around Europe, especially in the North Atlantic, and called for the United States and its allies to once again meet Russia at sea in a comprehensive manner. Foggo's article stated, among other things, that "the submarines of the Russian Federation are one of the most difficult threats the United States has faced. This threat is significant, and it is only growing in complexity and capacity."[1] Admiral Foggo is known as a naval leader who is not afraid to reach out to broader audiences with his thoughts and ideas. He is also sensitive to the fact that in the new contest between Russia and the United States, global public opinion is a battlefield too. In the wake of the close flyby of the USS *Donald Cook* in the Baltic Sea in April 2016, Foggo was quick to release images and video from the destroyer's bridge wing, which later were played on the evening news across the United States and Europe. As a result, the Russian flybys of the *Donald Cook* were seen by the public as both provocative and dangerous.

But Foggo was not the only one who decided to raise the alarm about the Russian navy in the Atlantic and elsewhere. In a speech in Washington in late 2015, Adm. Mark Ferguson, the commander of U.S. Naval Forces Europe, described Russia as building an "arc of steel from the Arctic to the Mediterranean" that would produce a "sea denial strategy focused on NATO maritime forces." Ferguson also pointed to the growing Russian ability to project naval power from this arc of steel, especially with Russian subsurface forces.[2] Non-Navy officers also worried about security in the North Atlantic. Gen. Philip Breedlove, USAF, who came into the job as commander of United States European Command and as NATO's supreme allied commander Europe just as Russia's new aggressiveness became apparent, publicly worried about NATO's ability to come across the North Atlantic during a crisis or while at war. Shortly after retiring, Breedlove laid out his case in public: "The unobstructed crossing of the Atlantic to fight a war on the land mass in Europe, I think, is a thing of the past. We need to think about our ability to defend our capability to reinforce Europe. . . . In a time and space of their choosing, they [Russia] can make things very tough for us, and we need to be able to ensure our ability to operate in those commons."

Under the leadership of Chief of Naval Operations John Richardson, the U.S. Navy also began to reassess its role in an increasingly competitive and contested maritime domain, and in the Navy's new "Design for Maintaining Maritime Superiority," Russia, along with China, were pointed to as being in strategic competition with the United States and using their navies to that end. "The Russian Navy is operating with a frequency and in areas not seen for almost two decades," the Navy's new document observed.[3] The alarm was raised by European navy leaders as well. While speaking at the Belgian embassy in London, the commander of NATO's Maritime Command in the United Kingdom, Royal Navy Vice Admiral Clive Johnstone, called Russia a "fierce and uncompromising competitor" in the maritime domain and said that NATO needed to gain "a new and deeper understanding of the North Atlantic and its relevance to Alliance security."[4] The United Kingdom's military chief, Air Chief Marshal Stuart Peach, also publicly warned specifically about the vulnerability of the submarine cables across the Atlantic during a public address at the United Kingdom's most respected think tank, the Royal United Services Institute. "Can you imagine a scenario where those cables are cut or disrupted, which would immediately and potentially catastrophically affect both our economy and other ways of living?"[5]

That Russia and its navy was once again featured in the key strategic documents of the U.S. Navy, and other services, is no small matter. The return of Russia to the firmament of potential U.S. near-peer military competitors is a break with nearly two decades of thinking that previously was dominated by nonstate threats, by China, and by, as a distant third and fourth, Iran and North Korea as regional military powers that potentially could threaten U.S. and allied interests with cruise and ballistic missiles. But as senior naval commanders on both sides of the Atlantic began to ring the alarm bell, most NATO nations, due to the long period after the Cold War characterized by underfunding, operations elsewhere, and the strategic assumption that great-power competition would never return to Europe, were unprepared to deal with a resurgent Russia navy in the North Atlantic. A 2016 public report, based on a war game with participants from NATO nations across the North Atlantic region, on the re-emerging importance of the GIUK gap and the North Atlantic from a respected Washington think tank found that "the alliance is nowhere near ready to respond quickly to undersea challenges." The same report went on to note that it would likely take NATO almost two months to marshal an effective ASW force for the North Atlantic during a crisis with Russia.[6] Another credible and unclassified report from the Center for Strategic and International Studies in Washington declared that "Russia is expanding its undersea operations as part of a broader strategy of coercion aimed at its neighbors, NATO, and the United States. . . .

The Russian navy's use of submarines to signal presence, reach, and power achieves an effect that is disproportionate to the resources committed. . . . NATO and partner navies do not currently possess the ability to quickly counter the Russian subsurface challenge in much of the North Atlantic."[7] The warnings on paper and in statements about the resurgent Russian navy and its submarine force were soon complemented by real-world examples. In the summer of 2017, the U.S. aircraft carrier the USS *George H.W. Bush* and its escorts attempted to track the improved *Kilo*-class boat the *Krasnodar* in the Mediterranean after it had fired Kalibr cruise missiles into Syria. The submarine had been tracked by both NATO and the U.S. Navy as it entered the Mediterranean, but after the missile launches the boat disappeared, as the U.S. carrier group struggled to keep up. U.S. Navy P-8s flying out of Italy soon joined the search. A week later the *Krasnodar* reappeared as it fired another volley of Kalibrs into Syria and then again evaded detection by the carrier and its escorts. With the apparently fruitless pursuit of the *Krasnodar* in the Mediterranean that would last for weeks, the increasing sophistication of the Russian submarine force and the atrophied ASW skills of the United States and its NATO allies came into stark view.[8]

But in the wake of the Ukraine crisis and the by then clear evidence of Russia's reawakened interest in contesting the control of the North Atlantic and other maritime domains, some of NATO's Atlantic nations began to take the first steps to restore their abilities to operate in the Atlantic against a sophisticated adversary.

Reclaiming the Tool Kit

The United Kingdom was rudely awakened to the challenge of growing Russian naval activity and challenge in the North Atlantic in November 2015 after a Russian submarine was detected off Faslane in Scotland, where the United Kingdom's four *Vanguard*-class SSBNs are based. This incident was no mere matter of Russian probing of British waters. A Russian ability to detect and track a British SSBN as it heads out on its deterrence patrol would jeopardize the credibility of the British nuclear deterrent, as it is the only means by which Britain can deliver a nuclear strike. Previously Britain had used its MPAs in tandem with its SSBNs as they headed out to sea in an effort to clear the water space from lurking foreign submarines and give the SSBNs a detection-free passage to the open ocean.[9] With the British MPAs scrapped, this was no longer an option. Britain instead called on French, Canadian, and U.S. MPAs to join the search for the suspected Russian submarine off the coast of Scotland. A similar event with a suspected Russian submarine near Faslane occurred in early 2016, and later the same year two *Akula*-class SSNs from the Northern Fleet were tracked through the Irish Sea by a British SSN as the Russian boats made their way to the Mediterranean from the Kola Peninsula.[10]

In the wake of the Faslane incident and the growing realization that Russia had returned to the North Atlantic, London quickly made the decision to restore its MPA fleet by buying nine P-8 Poseidons from the United States. The British effort to bring back its airborne ASW capability is remarkable in its ambition. Normally, buys of such a magnitude can take up to a decade from start to finish and would include a carefully laid-out requirements study, a complicated bidding process for the interested suppliers, and negotiations between the winner of the contract and potential local suppliers of subsystems or maintenance and upgrades once the jets arrived. In the British P-8 case, much of this was laid to the side, as London decided to procure essentially a set of U.S. Navy P-8s, complete with American sensors and other systems. The swiftness of the British process was helped along by the fact that the British military had kept a seed corn initiative, where British MPA crews could maintain some of their skills while flying as part of American and Australian MPA crews. Still, the reintroduction of MPAs into the British maritime inventory will not happen overnight. The British government expects to have all P-8s delivered and flying operations over the North Atlantic and elsewhere in 2021, barring any unforeseen delay.[11] This is still a full five years after the initial decision to restore the British MPA capability.

Norway, which is NATO's most northerly member in Europe, also began to make moves to strengthen its posture and capabilities for operations in the Barents Sea and the far North Atlantic. In late 2015 Norway announced that it would replace its aging P-3 Orions with five P-8 Poseidons and was looking for opportunities to deepen cooperation in MPA operations in the North Atlantic with both the United States and the United Kingdom. Norway also began to seriously consider a replacement for its old *Ula*-class submarines, which came into service in the early 1990s. A replacement buy planned earlier had been held off due to a shrinking defense budget and priorities elsewhere. At the same time, Oslo began to work with Washington on increasing the U.S. presence in the High North, which came to fruition with a decision late in the Obama administration to deploy a few hundred U.S. Marines on a rotational basis for exercises and training with Norwegian forces. This Marine Corps presence in Norway grew further to roughly seven hundred Marines during the first two years of the Donald J. Trump administration.[12] Inside NATO, Norway also became a vocal proponent for a new look at the maritime domain by the alliance and at the capabilities and command structures needed by NATO to operate at sea against a qualified opponent.[13]

And while Norway had some catching up to do with a resurgent Russian navy in the North Atlantic, the nation was not as far behind the power curve as others. Norway is a maritime and Atlantic nation at heart, and the waters surrounding the nation have always been considered an important national interest, be it economically through fishing and oil and gas extraction or as the great connection that provided Norway with access to the broader North Atlantic region. Norway's P-3s had also remained on duty in the Barents Sea during the 1990s and early 2000s, albeit with a shift toward fishery patrols, environmental monitoring, and stints in support of NATO out-of-area operations.[14]

As a result of the resurging Russian navy, Iceland also gained back some of the attention the tiny island nation enjoyed during much of the Cold War, as its importance as a dry-land outpost in the middle of the North Atlantic once again dawned on decision makers in Washington and at NATO headquarters in Brussels. Keflavik airbase, now a predominantly civilian airport, saw visits by U.S. Navy P-8s that occasionally would patrol the North Atlantic out of Iceland. But tweaks were required to enable the P-8 patrols out of Iceland. The P-3 hangars that had been kept up by the Iceland government were not a fit for the larger P-8s, and a new notch had to be cut in the front face of the hangar before it would accept the slightly taller tail of the Poseidon.[15] More upgrades to the infrastructure in and around Keflavik are in the offing, as part of the U.S. European Deterrence Initiative that provides funds for renewed U.S. military activities in Europe.[16]

But NATO's and America's return to Keflavik is not a small matter of simply showing up and turning the lights back on. Keflavik has moved on since the U.S. departure in 2008. Today, many of the old base buildings still stand but have been taken over by Icelandic start-up companies or serve as simple hotels. Still standing is the base theater, "Andrews," which used to show American Hollywood productions for U.S. service members and their families while stationed there, but its parking lot is being slowly but surely claimed by weeds that are piercing the asphalt in the parking lot. The main hangar is now used by Atlantic Studios, which assists with film and TV productions on Iceland, including the HBO smash hit *Game of Thrones*. But a small corner of the base is still a restricted area and is used for the occasional visits of jets from other NATO members. They are there to fly as part of NATO's air policing mission over Iceland, which was started after the American departure and which happens for a few weeks at a time about four or five times a year. Meanwhile, the civilian airport right next door to the airbase has aggressively expanded as commercial and tourist interest in the High North has grown. In 2016, Iceland welcomed some two million tourists and served as a hub for some six million transit passengers between Europe and North America.

But after the Americans left, the Icelanders decided to quietly keep a few lights on in the event that the security situation would once again change. The Iceland coast guard, an organization of roughly 250 personnel, which also has defense responsibilities in the tiny country without a military force, maintains and operates a chain of radar sites around the island, which provide a good air picture over an area about the size of central Europe. The coast guard has also invested in search-and-rescue helicopters to replace the ones removed by the United States in the early 2000s, which can be used to pluck military pilots out of the sea as well as fishermen. Iceland also pays for the upkeep and security of the corner of Keflavik being used for the NATO air policing mission.[17] Iceland will never be able to do what the United States and NATO can do in the maritime domain in terms of ASW and ISR, but it is determined to make it as easy as possible for its bigger allies to return to Iceland to do those jobs. And it seems as if that time has come. The P-8s will likely not come to Iceland on a permanent basis. They are far too valuable and there are far too few of them to completely dedicate even a few to patrolling around Iceland, but this does mean that American sub hunters will once again fly from Iceland and around the GIUK gap.

Other specks of land in the North Atlantic that had played a role in the alliance network of bases and sensors during the Cold War were also of interest. In 2015 the Norwegian military upgraded infrastructure and the Loran-C station on Jan Mayen island, which sits north of the GIUK gap.[18] In late 2017 the British government made the decision to reopen the radar station at RAF Saxa Vord on the Shetland Islands, a site that had been used during the Cold War to track Soviet aircraft coming out over the North Atlantic from the Kola Peninsula.[19]

The United States and NATO also began to rebuild some of the structures needed for more ambitious operations in the North Atlantic in the future. In the late summer of 2018 the U.S. Navy re-established the Second Fleet, which had been retired in 2011, to help prepare U.S. naval forces for operations in the North Atlantic, including the Norwegian and Barents Seas. In his speech during the establishment ceremony of second fleet in Norfolk, Virginia, Chief of Naval Operations Adm. John Richardson charged the Second Fleet with gaining a "central role in pioneering new and experimental concepts of operations" in close concert with U.S. allies and partners around the North Atlantic.[20] During the same summer NATO decided at its summit in Brussels to create a Joint Forces Command–Atlantic, to be headquartered in Norfolk, Virginia, the same location where ACLANT once stood. The new command is intended to be the nucleus for a new Atlantic-focused NATO command structure. In the late fall of the same year, NATO returned to the North Atlantic and north of the GIUK gap in force with the exercise Trident Juncture, with an exercise area that stretched from Iceland to northern Norway. The exercise included the carrier *USS Harry Truman*, a sight not seen in the far North Atlantic in more than two decades.

Reinforcing Europe across the Atlantic Once Again

With NATO once again focusing on deterring and defending against aggression aimed at the alliance's new members, the key role played by the North Atlantic in defending NATO's new and exposed eastern members began to slowly but surely dawn on the alliance. During its 2016 summit in Warsaw, NATO decided to deploy battalion-sized groups to Estonia, Latvia, Lithuania, and Poland. These groups are led by the United Kingdom, Canada, Germany, and the United States, and additional companies or platoons are contributed to them by other allies, such as Norway, Italy, the Netherlands, and Croatia. These enhanced Forward Presence (eFP) groups, in NATO parlance, are not intended to halt a Russian advance in case of war. Instead they are signals of NATO's political determination, as Russian forces will inevitably have to engage the eFP troops (and by extension many of the members of the alliance) if Russia intends to seize the Baltic states. In many ways, the battalion-sized eFP groups in Europe's east, made up of Americans, Britons, Germans, French, Dutch, Croats, Canadians, Norwegians, Danes, and others, serve the same purpose as the allied brigade in Berlin during the Cold War: they are placed in harm's way to raise the cost of aggression for Moscow and commit the allied nations to war if they are ever attacked. In a crisis these battalion groups would be aided by NATO's rapid reaction force, or the Very high-readiness Joint Task Force (VJTF), which would offer a brigade's worth of combat power on the relatively short notice of about a week or two. NATO's heaviest and decisive punch would be delivered by follow-on forces from the rest of Europe, but in particular the United States, which would have to move to the region. In short, NATO's approach to deterring Russia and defending against aggression in northeastern Europe is based on a reinforcement model, in a fashion that is even more pronounced than during the Cold War due to the simple fact that the U.S. military footprint in Europe today is far smaller and lighter than it was during the Cold War.[21] At the height of the Cold War America maintained some 400,000 troops in Europe; today that figure stands at around 60,000. Even though an uptick in the permanent U.S. forward presence in Europe in the future is possible, it is clear that it will remain smaller than during the Cold War and will have to be reinforced in the case of war or crisis. And here, as it did during both world wars and the Cold War, the North Atlantic will need to provide the bridge between Europe and North America.

American reinforcements across the Atlantic will be, just as during the world wars and the Cold War, a fundamentally maritime undertaking. While the United States maintains the largest fleet of airlifters in the world, it is nowhere close to being able to do the job of moving a substantial combat force across the ocean. In carrying capacity, one journey across the Atlantic of a fast sealift ship equals roughly 130 flights with the U.S. Air Force's C-5 galaxy heavy lift aircraft.[22] To move a fighting force across the sea is therefore no small feat. Once American forces and equipment destined for northern Europe have crossed the Atlantic, they will have to be funneled through a limited number of European ports that have the capacity to quickly offload the ships, stage the equipment and personnel into units, move the assembled units toward the east, and arrange for the handover to the unit commanders and higher headquarters.[23] Ammunition shipments must be specially handled by particular ports equipped and designed to accommodate the storage and transfer of them, with only Nordenham in Germany currently being able to do that job.[24] Given the new Russian emphasis on submarines that can launch land-attack missiles, these ports (and the airports that would be used to fly in additional U.S. personnel) are potential targets that have little in the way of defense or physical hardening. They could therefore be destroyed or reduced in capacity in a Russian bid to deny U.S. access to the European continent. They could also be subject to sabotage by infiltrators, a worry that also was present during the Cold War.

And once across the Atlantic, American and allied units will have an insatiable appetite for supplies, ranging from fuel and food to spare parts and ammunition. In recent high-intensity operations, American brigades have been known to consume more than 22,000 gallons of fuel in a twenty-four-hour period. The U.S. Army has calculated that an M1A1 Abrams main battle tank requires roughly forty 120-mm tank rounds for every day of fighting. Weighing in at nearly 50 pounds each, it means that an armor battalion of around fifty-five tanks require 2,200 rounds for its main guns per day, weighing in at around 50 tons. And a modern ground force requires a full range of munitions, from small arms rounds and grenades to artillery rockets, guided antitank missiles, and mines.[25] Also, modern fighting forces, reliant on miniaturized high-tech equipment, have added an additional logistics requirement: mountains of batteries. The United States has conducted these types of logistical feats repeatedly since the end of World War II. During Desert Storm, for example, some five hundred ships were used to bring, among other things, 41,000 containers with supplies, 12,000 vehicles, and 400,000 tons of ammunition into Saudi Arabia to support the liberation of Kuwait.[26] These undertakings, however, have not been under conditions where the enemy has sought to strike and destroy the vital nodes needed to accept these mountains of equipment and consumables.

The logistics tail across the sea can to some degree be reduced by forward-basing equipment and supplies, which drove part of the reason for basing U.S. forces in Europe during the Cold War. And the Marine Corps has maintained a brigade's worth of equipment and supplies in caves in northern Norway since that time. The United States is also putting supplies and equipment into new facilities in eastern Europe and Germany. While those efforts are helpful and mean that some units could fly their personnel into country where they could join up with their equipment, it does not obviate the need for bringing transports across the sea containing vehicles, machines, parts, ammunition, and fuel. As the U.S. Army and Marine Corps are stepping up exercises in Europe to once again get to know the terrain, working with American allies, and sending a message of deterrence to Russia, the U.S. military has also begun exercising the muscle memory of bringing reinforcements across the Atlantic, an effort that has not been undertaken for more than twenty years. It is not an undertaking that can just be drilled on paper or at the staff level. It must be done under real-world conditions. "You have to do it physically. . . . There is no way you can discover all the friction until you actually have people and stuff moving," explained the commander of U.S. Army Europe, Lt. Gen. Ben Hodges, in an interview to *Inside Defense*.[27]

And just like the Cold War, it is an arduous process that begins by meticulous packing and preparation at home bases in the United States, with the equipment then brought forward to ports along the U.S. east coast and the Gulf of Mexico. And in early 2017 the United States began rehearsing once again what it had not done for more than two decades: lifting a U.S. Army brigade across the Atlantic with all of its personnel, equipment, vehicles, spare parts, and other items. First out of the gate was the 10th Combat Aviation Brigade from the 10th Mountain Division, which arrived via ships to Antwerp in Belgium and Bremerhaven in February 2017.[28] This first major sealift of U.S. forces back to Europe in more than two decades included some 2,000 soldiers, more than 100 helicopters, and some 350 ground vehicles. In the coming years, the U.S. Army is expected to start bringing division-sized forces across the North Atlantic for exercises and training in Europe.[29] To support all this, the U.S.

military sealift command (MSC) is also busier in the Atlantic than it has been for decades. Between 2016 and 2017, MSC transports of ordnance across the Atlantic more than doubled, while the moving of critical parts more than tripled.[30]

CHAPTER 12

The Contested North Atlantic in the Twenty-First Century

THE RETURN OF FRICTION and competition to the North Atlantic has already generated an initial response from the United States and some of its NATO allies in Europe. American P-8s are once again occasionally flying from Keflavik in Iceland, and in 2017 NATO held its annual ASW exercise, Dynamic Mongoose, in the North Atlantic with a new level of intensity. And in 2018 NATO returned to the North Atlantic for the major exercise Trident Juncture, and also began to build a small Atlantic-focused command based in Norfolk, Virginia. At the same time, the U.S. Navy stood the Second Fleet back up in order to be better prepared for naval operations in the North Atlantic region. Norway has decided to buy P-8 Poseidons to replace its aging P-3 fleet, while the United Kingdom has announced that it too will buy P-8s to restore the MPA fleet that was removed from service in 2010. The U.S. Navy has also launched an "Ocean Task Force" to expand ocean science relevant to naval operations, particularly those aspects, such as updated sea floor mapping, that are helpful to ASW.[1] NATO, meanwhile, is sponsoring science and technology investigations in the North Atlantic to help alliance navies better understand the subsurface conditions in this once again vital maritime domain.[2] All the above are useful and necessary efforts, but they are just a start and are by themselves not sufficient over the long term to ensure that the Atlantic remains open as the communications, trade, and military reinforcements maritime superhighway between North America and Europe in peace, crisis, and war.

What Remains the Same

While the twenty-first-century competition in the North Atlantic should not be thought of as a simple rerun of the three battles fought during World War I, World War II, and the Cold War, history can be instructive in bringing to light the enduring strategic factors in this maritime domain. Twentieth-century military events in and around the North Atlantic show that to win the competition, one needs a strategy; that the contest will be technology intensive; that it will be a long-term endeavor; that it will at times lead to tense naval interactions that must be managed; that the challenge must be correctly analyzed and diagnosed; and that access to key land features will be crucial.

Just as during the world wars and the Cold War, military conflict between Russia and NATO in the twenty-first century is unlikely to start in the North Atlantic. It is also hard to imagine that a conflict would be sparked over some issue that is directly related to the maritime domain in the region. Instead, the North Atlantic, just as during the world wars and the Cold War, would be an important supporting theater to military events that have been begun to progress elsewhere, perhaps on a piece of territory far from the sea. But by operating in the far North Atlantic and attacking vital ports, airfields, and command-and-control centers with long-range cruise missiles in an effort to shut down the reinforcement efforts of the United States and NATO, or deter the alliance from taking action to begin with, Russia could stop a NATO response to Russian aggression before it even began. This is why a crisis over the Baltic states or somewhere in eastern Europe could quickly have reverberations in the North Atlantic. So far, the United States and its European allies have been too tightly focused on addressing relatively small regions of potential conflict between Russia and NATO and have failed to think in broader strategic terms about how a potential conflict could escalate and spread far away from its original source, including to the North Atlantic and its nearby seas.

Other key factors that were highlighted during the previous battle for the Atlantic remain relevant as well. The competition is sure to be technology intensive, with both Russia and the NATO allies around the North Atlantic investing in high-end submarines and the platforms and networks to hunt them. The emerging competition will also require close teamwork among the allies, just like during the previous three competitions.

And What Has Changed

It is also important to be clear what is different about the emerging maritime contest in the North Atlantic in the twenty-first century compared to the naval competitions during the past century. First, the Russian navy is not the Soviet navy in terms of numbers or operational reach, but that does not mean that the Russian navy is not a serious operator in the far North Atlantic. And while there is a stated Russian ambition to return to the world's oceans as a blue-water navy, economic and technological limitations suggest that Russia will likely only be able to put to sea a green-water force over the coming decades. Still, this is a fair accomplishment given the doldrums that the Russian navy, including the Northern Fleet, sank into after the Cold War. In spite of its shortcomings and stumbles, the work that the Russian navy has done, encouraged and funded by Putin, to claw its way back to relevance is impressive. Indeed, the current and coming Russian submarine force is far more skilled and agile than the one that put to sea during the Cold War. The new Russian navy is focusing on smaller numbers of submarines and ships, but all of them of higher quality and leveraging the increasing availability of accurate long-range weapons against both ships and targets ashore. This approach does not make the Russian navy a global competitor to the U.S. Navy, but it presents a real challenge for

NATO and the United States in the North Atlantic and the other seas surrounding Europe and in particular north of the GIUK gap. The Russian challenge in the far North Atlantic is further complicated for the United States and its European allies by an increasingly turbulent global security landscape, evolving technologies, and changing geopolitics of the North Atlantic region.

A Busier World for Navies and a Declining West

The world as a whole is more militarily competitive than it was two or three decades ago, and this will have implications for the defense of the North Atlantic. The U.S. Navy, for example, will not be able to singularly focus on the North Atlantic domain, but will be forced to carefully balance demand signals for its presence in both the Atlantic and the Pacific. Achieving this balance between an Atlantic and Pacific focus will not necessarily become any easier as the world progresses further into the age of great-power competition, a security environment that is sketched out in the Trump administration's National Defense Strategy, a strategy document released in late 2017 that has found bipartisan support in the U.S. national security community.[3] It is also likely that the Middle East will remain turbulent for decades to come and will require a U.S. offshore presence for, among other things, strikes, raids, and noncombatant evacuations. This is a tall order for a U.S. Navy that is already coming off two decades of operations around the globe as part of the global war on terrorism and is already struggling to meet needs for both a forward presence at world hot spots and catching up on its maintenance requirements.[4] Indeed, the wear and tear on the U.S. Navy's ships and their crews are beginning to show in places, not least with the 2017 string of at-sea mishaps and collisions that ended up killing sailors. More often than not these incidents have been attributed in part to tired and overworked crews and lack of training due to the high operational tempo.[5] Bearing this in mind, there is little to suggest that the U.S. Navy's permanent forward presence or posture in Europe will be radically larger in the coming years, leaving it with four destroyers based in Spain, primarily for missile defense tasks, and a command ship in Italy. The U.S. submarine force, which was crucial during the contest in the North Atlantic during the Cold War, may stand as a practical example of this global conundrum for the U.S. Navy. The U.S. submarine force today is certainly smaller than it was during the Cold War, and today it must also contend with a rapidly growing Chinese submarine force, which may not be as sophisticated as the Russian one but is highly active and operating farther and farther from Chinese shores.[6] Indeed, the majority of U.S. attack submarines are today based in such a way as to be able to swiftly access the Pacific rather than the Atlantic.[7]

This reality of competing demands in the maritime domain is to some degree mirrored among America's European allies. They too face maritime threats of various natures from several points of the compass. While northern European nations such as Norway, Britain, the Netherlands, Denmark, and Germany are increasingly focused on the North Atlantic and other northern waters and the kinds of maritime capacities and capabilities needed there, that is not the case for southern Europe. In the Mediterranean, softer maritime security issues still reign supreme, with the refugee flows out of the Middle East often taking the sea route to countries such as Italy and Greece. Not infrequently the journey across the Mediterranean ends with tragic results and is accompanied with heartbreaking images splayed on the front pages of newspapers. No wonder at all that southern European nations are therefore more keen to build out their coast guards rather than upgrade their navies. The current turbulence in North Africa and the Middle East, which drives the insecurity in the Mediterranean, may subside in the coming years, but long-term demographics and state fragility in Africa and the Middle East suggest continued turbulence and therefore that the Mediterranean will remain an active zone for maritime security efforts. And even if a maritime division of labor of sorts is emerging within Europe, in which the southerners tend to the human tragedies playing out in the Mediterranean and the northerners are turning toward the Russian challenge at sea, tensions remain within the alliance on what NATO's role should be in each maritime corner of Europe.

Adding to this tension are the many relatively new NATO allies, such as the Czech Republic, Hungary, and Slovakia, who are all landlocked and therefore understandably more focused on the land and air domains of defense. And the geographic mindset of NATO's members should not be underestimated as a driving factor in their worldview. NATO's founding members, including the United States, the United Kingdom, the Netherlands, France, Portugal, Norway, and Denmark, were all to a large degree Atlantic nations. NATO members who joined later during the Cold War may not have been Atlantic-facing, but they were all certainly maritime nations, with members of this group including Spain, Greece, and Turkey. As NATO enlarged toward the east in the post–Cold War era, the alliance's mindset also shifted from its Atlantic and maritime roots. The tensions, different orientations, and domain mindsets across the alliance are no mere curiosity, as NATO moves only after consensus has been achieved among all allies.

This broadly more competitive global landscape combines with the relative decline in Western military power that has quietly progressed over the past two decades. The United States and its European NATO allies simply do not have the same resources to throw at the challenges in the North Atlantic today as they did during the Cold War. MPAs, for example, are a so-called high-demand/low-density asset today, meaning that they are called on to perform a large number of missions on a routine basis; but there are only relatively few aircraft to go around to cover the demand for them. Submarine fleets within NATO, meanwhile, have been on a steep decline since the Cold War. While real defense spending has certainly declined, the ever increasing cost of advanced military systems has played a real role too in depressing the number of ships, submarines, and aircraft.[8] In the 1980s, a P-3 Orion cost roughly $80 million in today's dollars. The new P-8 Poseidon, currently being brought into service by the U.S. Navy, Norway, and the United Kingdom, costs around $125 million, an increase of more than 50 percent. Conventional submarines, like those operated by many of NATO's members, certainly cost less than the nuclear-powered boats operated by the United States, France, and Britain, but that does not mean that they are cheap. Today's advanced air-independent propulsion submarines, such as the German-made *212* class, can reach $650 million per boat, before considering in-service and later modernization costs. This is a significant investment for many small nations with defense budgets that do not reach $10 billion per year. In contrast, the *212*'s forerunner, the *209*, could be had for a little more than half of that cost. Measuring the real costs of complex naval systems that are decades in the making is of course an inexact proposition, but the trend is unmistakable. Costs are rising with each new turn in the military-technology cycle. Of course, the advanced MPAs and submarines of today are far more capable, effective, and efficient than their Cold War forebears in terms of their ability to surveil their surroundings, track targets, process and fuse data, and attack targets with a high degree of precision. This offsets the lower number of systems in the inventory to some degree, but at some point quantity becomes a quality of its own. An MPA or an ASW frigate can only be in one place at any given time, and smaller numbers means that a force will quickly become stretched. ASW is also not likely to change in its most basic character, meaning that it will remain a tedious and long-term task out at sea. With smaller numbers of aircraft and ships, however sophisticated, this means that being able to sustain an ASW operation over a longer period of time becomes a real issue. And even if nearly unlimited funding were made available to northern European navies, it is doubtful they would be able to restore the capacities found in the navies of Europe just a decade or so ago. The Netherlands gave up its MPAs and Denmark walked away from its submarine force in the early 2000s, and just buying new aircraft and submarines would not bring these capabilities back. It would require several decades of reinvestment in infrastructure, training, personnel, rebuilding the knowledge base, and even re-establishing a service culture for Denmark and the Netherlands to credibly operate these systems again. That is a daunting feat to imagine, even if the political will were there.

The Changing Geopolitics of the North Atlantic

Today the North Atlantic region is a more international space than ever before, and this trend is likely to continue given the trajectory of the global economy and the rise in prominence and power of non-Atlantic nations. During the Cold War the North Atlantic was certainly busy with economic and military activity, directly driven by the nations around that maritime domain. Today and into the future, though, the North Atlantic region is also an object of interest for emerging global powers that hail from places far away from the core North Atlantic community. China has a growing commercial interest in the North Atlantic region, with investments in natural resource extraction and infrastructure around Iceland, Greenland, northern Norway, and the Kola Peninsula. China also has a budding scientific center on Svalbard, which was first opened in 2013.

China's interest in the far North Atlantic derives from Beijing's Arctic ambitions. The region around the top of the world is viewed as being able to help solve two of China's most serious challenges: access to energy and other natural resources that could help fuel China's continuing economic rise, and, through the use of the northern sea route, liberate Chinese shipping from the maritime choke point at the Strait of Malacca or from reliance on the Suez Canal, which brings Chinese trade through the politically precarious and turbulent Middle East.[9] China's interests in the North Atlantic and the Arctic may be primarily commercial rather than driven by national security, but commercial interests have a way of attracting military power as well, a fact so clearly laid out by America's own Alfred Thayer Mahan in *The Influence of Sea Power upon History*. Indeed, China's People's Liberation Army Navy was first spotted in northern waters in 2015, when a small task force made up of a destroyer, a frigate, and a supply ship passed through the English Channel, entered the North Sea, and then proceeded through the Kiel Canal into the Baltic Sea for port visits in Denmark, Finland, and Poland. The 2015 visit to the North Sea and the Baltic by the Chinese navy was a good will tour with a limited number of ships, but the Chinese navy returned to the Baltic during the summer of 2017 for a joint exercise with Russia's Baltic

Sea Fleet, which included combat elements. These small but growing Chinese maritime advances into northern waters could very well be a harbinger for a more regular, if occasional, naval presence of China in the broader North Atlantic region.[10]

The emerging Chinese interest in the nations and maritime domain in the North Atlantic region has already led to friction at times, although it has been quietly managed by the nations involved. A Chinese businessman with links to the Communist Party sought to buy big tracts of land in both Iceland and Svalbard but was rebuffed in both places by the Icelandic and Norwegian governments over concerns that the purchases were thinly veiled attempts by the Chinese government to gain physical footholds on sensitive pieces of land.[11] And in 2014 the Norwegian government denied permission for the Chinese government to add a radar system to its scientific outpost on Svalbard over concerns that the system was intended for intelligence gathering rather than studying solar winds, as the Chinese claimed.[12] Chinese construction companies have also signaled an interest in expanding the few airports that connect Greenland to the outside world.[13] Indeed, in 2018 a Chinese company aggressively bid to build three new airports on Greenland, which could have had implications for the U.S. presence on Greenland and its missile warning and space surveillance systems there. Ultimately, the Danish government moved to disqualify the Chinese company from the bid as a favor to the United States.[14]

Farther south, China has also taken an interest in the Azores, a collection of islands in the mid-Atlantic under the control of Portugal that served as a crucial base for U.S. and allied ASW patrols during both World War II and the Cold War. In 2015 Chinese premier Wen Jibao visited the islands on his way home from a jaunt through several Latin American countries. While he was there, the Chinese indicated an interest in both the airfield and the main port on the Azores and their potential as an ideally placed logistics hub in the southern North Atlantic, a location within easy reach of the Mediterranean.[15] And China seems to be moving forward with this interest in mind. In 2016 the Chinese government approached

Portugal to discuss opportunities for joint deep-sea research using the Azores as a base.[16]

That China's presence in the North Atlantic is growing does not automatically mean that the United States and NATO will have to contend with yet one more potential adversary in the North Atlantic maritime domain. But the addition of another powerful player, albeit from the other side of the world, further complicates the geopolitics of the region and adds yet another set of actors that will require watching. And it should not be lost on strategists that China and Russia are increasingly fellow travelers in their attempts to tilt the world order in their favor and call into question the global rules of the road guarded by the United States and its allies. At a strategic level there are stunning similarities between Russia's aggression in Ukraine and Moscow's continued assertiveness against NATO's members on the one hand, and China's efforts to establish pre-eminence in the South China Sea on the other.

Political and naval leaders on both sides of the North Atlantic must also be mindful that there may yet be more major geopolitical movements in the North Atlantic in the coming years, with independent microstates emerging out of Europe's larger countries, which could have a major impact on NATO's toeholds in the maritime domain. The British popular vote to leave the EU in 2016 has first and foremost caused sharp worries about the future of a cohesive and strong Europe; but further downstream one of the effects of a British departure from the EU could be that Scotland (a part of the United Kingdom that voted overwhelmingly to remain in the EU) will make good on its move for independence and attempt to rejoin the EU as an independent nation.

And Scotland may not be the only nation in the North Atlantic region seeking to assert its own identity in the coming years. Greenland may once again seek independence from Denmark at some point in the future, which would in turn raise questions about the use of bases and radars on Greenland that served such crucial purposes during the Cold War and are once again today relevant to both Denmark and the United States.[17] The Faroe Islands, which sit between Iceland, Britain, and Norway in the Atlantic Ocean, may also seek independence from Denmark in the coming years. The Faroe Islands do not have much in terms of military infrastructure today, but during the Cold War they were home to a NATO early warning system and a Loran C navigation system. Faroe real estate may become valuable again in the new competition between NATO and Russia's Northern Fleet.

If the Atlantic nations of Europe let their pieces of land in the North Atlantic slip away, it is far from certain that the newly independent micronations would remain in the exclusive orbits of NATO and Europe.[18] In the case of Scotland, its independence would have direct implications for Britain's submarine-based nuclear deterrent, which is currently based in Scotland.[19] The basing would have to be moved, placing them farther away from the open waters of the North Atlantic. The United Kingdom's resurrected maritime patrol capability is also intended to be based in Scotland, close to its primary operating environment in the North Atlantic. They would also have to move farther south if Scotland opted for independence in an attempt to re-enter the EU. Scottish nationalists are certainly international and European in outlook, but they have always maintained ambiguous feelings about the role and use of military power. It is far from a foregone conclusion that a Scotland outside of the United Kingdom would like to lend itself as a base for patrols in and over the North Atlantic. In essence, Scottish independence could very well mean that NATO would lose control over one of the gates of the GIUK gap and an ideal basing area for operations north of there.

These new states may also be attracted to powers far from the North Atlantic, including both Russia and China. Russia is today the Faroe Islands' largest trading partner after Russian countersanctions against the EU (a response to the EU sanctions on Russia after the annexation of Crimea) halted the export of European seafood products to Russia.[20] The Faroe Islands stepped in to fill the gap, and its fisheries are enjoying a boom in profits supplying salmon and cod to Russian markets.[21] Likewise, China is eyeing mining prospects in

Greenland, where climate change is opening areas that contain rare earth minerals crucial in the production of key technologies such as clean energy systems, electronics, and radars.

These are commercial interests and ties rather than political and military ones, but in the case of China and Russia the state plays a heavy hand in major industries and investments abroad. And commercial linkages are ties that truly bind and will give decision makers pause in deciding where to turn and what to do during a crisis or conflict.[22] Indeed, commercial links and trade dependencies have already influenced the dynamics between Russian and Europe ashore. Maintaining Europe's sanctions against Russia, imposed after the annexation of Crimea in 2014, have proven a continuing struggle, as several European nations are impacted by the loss of trade with Russia.[23]

And even countries and areas that are safely ensconced in NATO or one of the Atlantic nation-states are experiencing increased commercial and political interest from non-NATO and non-Atlantic powers. In recent years Beijing has noticed Iceland and its strategic location in the middle of the North Atlantic. China is one of the largest investors in Iceland, and Beijing maintains the largest embassy on the island, prominently placed in downtown Reykjavik. Its dark brick structure stands out against the more modest and mostly wood structures of the rest of Reykjavik. The size of the embassy is to some extent explained by the fact that all Chinese personnel are required to live at the embassy instead of out in town, which is the practice for most other foreign missions. Still, the outsized size of the embassy in Reykjavik makes it clear that China is serious about keeping an eye on the region and on Chinese commercial and strategic interests at the top of the world.

Svalbard is another key piece of terrain in the changing North Atlantic. The four-hundred-mile stretch between Svalbard and the Norwegian mainland provides a first choke point that the Russian Northern Fleet must pass to exit the Barents Sea and pass into the Norwegian Sea and on to the Atlantic. Svalbard is one of the remotest places in the world, but it is not without civilization and creature comforts. Longyearbyen, Svalbard's major town and named after the American coal mine magnate John Longyear, is strikingly modern and connected. Its population of roughly two thousand enjoys many of the social services offered to their Norwegian compatriots on the mainland. The town has, among other things, a modest shopping center, a medical clinic, regular distribution of mail, and a small police force. It even has a small college focused on Arctic scientific studies, which draws students from not only Norway but from across Europe. Cell phone service is excellent, and the Internet is surprisingly fast. Longyearbyen residents can also avail themselves of cheap tickets to and from Norway on regular flights that come in and out on a daily basis. None of this comes cheap, of course. To essentially project the Norwegian welfare state into the far reaches of the Arctic costs Norwegian taxpayers considerable sums in subsidies each year. The logic behind it, however, has little to do with notions of fairness or social democracy. Instead, it is about sovereignty. Extending the reach of the Norwegian state to Svalbard means less risk of a foreign actor (government or private) moving in to offer desirable goods and services and thereby potentially opening a rift, politically, socially, and economically, between Svalbard and the Norwegian mainland.

Russia has a presence on Svalbard, too, and is a party to the Svalbard Treaty, which gives sovereignty over the collection of islands to Norway but provides commercial and scientific access to all signatory nations. The Soviet Union established a permanent presence of sorts on the islands through two coal mining operations and adjacent company towns called Barentsburg and Pyramiden (the Pyramid). Today, only Barentsburg remains, where a few hundred Russians keep the coal mine going in a bleak existence with few of the creature comforts provided by Norway at Longyearbyen. The town and coal mine at Pyramiden folded in 1997, with the then weak Russian state unable to keep subsidizing its existence. A Russian plane crash in the area that killed some one hundred of its inhabitants served as the final death knell for the town.[24] The Russian coal mines on Svalbard have rarely if ever turned a profit, and that goes for the Norwegian coal mining efforts on Svalbard as well. But that is beside the point in the world's northernmost permanent settlement. The rationale for the mines is to provide a reason for that permanent presence on Svalbard, and Russia is not prepared to abandon any of its infrastructure on the island. To do that would mean losing the foothold on Svalbard forever.

Russian interest in Svalbard is once again increasing. In April 2015, Russia's deputy prime minister Dmitry Rogozin paid an unannounced visit to Svalbard, and while there he tweeted that the Arctic is a "Russian mecca." The visit turned into a row between Norway and Russia, since Rogozin is on the European sanctions list and is therefore not allowed to travel to EU nations or those that adhere to the EU's standards, as is the case with Norway.[25] As deputy prime minister, Rogozin has Arctic policy within his portfolio. Rogozin is also well known to NATO. Before becoming deputy prime minister, he served as Russia's NATO ambassador in Brussels, where he became notorious for lambasting NATO for taking on new members from the former Soviet Union and for its plans for extending U.S. missile defenses into Europe.[26]

Svalbard is also gaining in strategic importance in this age of space-based sensing, mapping, and science. Svalbard is ideally located at the top of the world as a platform to receive data downloads from orbiting satellites. No place on earth sees as many satellite orbits overhead each day. Svalbard today plays host to a large farm of satellite receivers, located on a ridge close to Longyearbyen; the receivers are all protected by sturdy white, round covers, making them seem like a field of mushrooms in the snow. The receivers are all run by the company Svalsat (owned by the Norwegian defense company Kongsberg), which provides data downloads for a host of clients, including NASA and the European Space Agency. Norway categorically states that none of the satellite receivers are used for military purposes, but Russia will sometimes suggest that it is not so sure. Regardless, the satellite receivers have certainly gained the attention of foreign powers; in 2007 and 2008 some of NASA's receivers on Svalbard were the subject of cyberattacks originating in China.[27]

The fishing grounds around Svalbard are also taking on a strategic quality in their own way. Here Norwegian, Russian, and European fishing fleets trawl for cod, mackerel, and other fish. While it has always been big business (according to the Norwegian coast guard a single haul of fish from a large trawler could be valued at up to a million dollars), it is now becoming a matter of national interest for Russia. As Europe imposed sanctions on Russia in the wake of the Ukraine crisis, Moscow applied countersanctions on European goods, primarily agricultural products. With less fish entering Russia from Norway, Finland, and other European fishery states, the Russian fishing fleets must provide enough of a haul to fill Russian grocery stores with fish at reasonable prices. The Putin regime's legitimacy partially rests on the appearance of its providing ever-growing prosperity to average Russians. Part of this narrative would fall away if middle-class Russians faced empty seafood counters at their neighborhood grocery stores. But the need for good catches goes beyond public popularity of the state. Russia's National Security Strategy from 2015 identifies self-sufficiency as a strategic objective for Moscow, and the associated "Food Doctrine" calls for 80 percent of Russia's seafood to be caught and processed by domestic means.[28]

Hybrid and Cyber Warfare in the North Atlantic

When thinking about the future of security in the North Atlantic, one must also consider that this is an age of hybrid warfare, where all elements of national power can be weaponized to advance hostile intent or create confusion or political division that will cause government and military decision making to grind to a halt. The operating environment in the North Atlantic and the role the North Atlantic plays in European security mean that hard and high-end military power will surely remain the key factor in the emerging contest between Russia and the United States and its NATO allies, but it should not be taken as a reason to discount the threat of hybrid warfare in the maritime domain. Indeed, Russian aggression and assertiveness against Ukraine, Georgia, and the Baltic states have all included hybrid elements in the maritime domain. For example, Russian sailors scuttled the aging *Kara*-class cruiser *Ochakov* in the inlet to Sevastopol in an attempt to stop the Ukrainian navy's access to the Black Sea during the early days of the Crimea annexation.[29] The late November 2018 incident in the maritime domain in the Sea of Azov, when Russian warships seized three Ukrainian naval vessels and their crews, proved to be another flashpoint in the ongoing struggle between Ukraine and Russia. Russian frigates also harassed the ships laying an electricity cable across the Baltic Sea between Sweden and Lithuania in 2015, to the point that the cable layers had to stop their work and leave the area. And in December 2007 the Northern Fleet's aircraft carrier *Kuznetsov* willfully conducted exercise maneuvers in such a way that flights to and from Norwegian oil and gas offshore installations had to be halted for a prolonged period of time.[30] With this recent history in mind, it is quite possible that maritime hybrid warfare, as well as marinized cyberthreats, could very well come the North Atlantic.[31]

Potential weak points that could be exploited in the North Atlantic region abound. The increased levels of economic activity in the North Atlantic region also brought new needs for communications, and a number of submarine cables were installed during the 1990s and into the 2000s between the United Kingdom and Iceland, Iceland and Greenland, and Norway and Svalbard. A network of submarine cables were also installed in the North Sea to assist the communications among the many oil platforms in the region and between Norway and the United Kingdom.[32] For example, Tampnet, a company operated out of Norway, provides fiber-optic communications to more than 240 oil and gas platforms and exploration rigs in the North Sea and provides shore connections to the United Kingdom, Norway, Denmark, and Germany. A physical or cyber disruption in that system would cause real economic harm and would send a strong political message, albeit without attribution, from Moscow. Indeed, Norway's state-owned oil and gas company, along with other energy-related businesses, was the object of a widespread and invasive cyber intrusion in the late summer of 2014.[33] And Russia's interests in the infrastructure on the North Atlantic seabed have increased in recent years, as discussed previously.

Russian special forces also practiced the seizing of oil platforms in the Arctic in 2014, citing the need to prepare for possible terror attacks against Russia's energy infrastructure in the region.[34] This is curious since the region is one of the least accessible in the world, and international terror organizations have not shown any interest in that particular locale. But the Russian special forces drills are of course just as applicable if Russia one day chose to seize an oil platform or another piece of infrastructure belonging to another nation in the North Atlantic region.

Russian disinformation operations aimed at maritime activities and naval operations must also be considered a real possibility. Disinformation operations are a Russian strength, and Moscow's efforts have ranged from stirring up discontent among Russian minority populations abroad to a serious and far-reaching operation to tilt the 2016 U.S. presidential election. And there are already indications that Russia has sought to use information campaigns to call into question the effectiveness of U.S. and NATO naval capabilities and their ability to operate in the seas close to Russia. The close encounter between the USS *Donald Cook* and a Russian Su-24 aircraft in the Black Sea in April 2014, which garnered considerable media attention, was quickly followed up by social media posts by a purported sailor (with clear signs that a nonnative English speaker was composing the text) from the USS *Donald Cook*'s crew who claimed that the Su-24 had knocked out the ship's computer systems and the Aegis radar with its electric warfare suite, effectively leaving the *Donald Cook* dead in the water. This story was later advanced by Russian government-supported media outlets, such as Sputnik and Russia Today, and later spread to European and U.S. news outlets, including Fox News.[35] Save for the actual close encounter between the *Donald Cook* and the Su-24, the story was entirely false, but the U.S. Navy never responded with an effective countermessage. Similar disinformation campaigns in the future could do real damage to the credibility of NATO and the United States at sea.[36]

Technology-Disrupting ASW

Along with the more competitive global landscape and a geopolitically changing North Atlantic come changes in naval and military technology. The previous battles of the Atlantic were resource- and technology-intensive, and the last one during the Cold War was based on the scattering of emplaced sensors over vast distances that fed a network of large and expensive platforms such as surface ships, submarines, and MPAs that could detect, track, and attack hostile submarines and warships. Indeed, the previous competitions over control of the North Atlantic to a large degree drove the development of ASW-related technologies. The first reaction of the Atlantic NATO nations in the emerging competition with Russia has been to seek to rebuild parts of the Cold War system of systems (the combination of bases, technologies, platforms, and relationships) in the Atlantic. Witness the British and Norwegian procurement of P-8 Poseidons from the United States and the U.S. rotational presence of MPAs at Keflavik. But this approach is only part of the solution and will require further investments.

For example, the SOSUS passive sonar arrays, and part of the IUSS, that proved so effective during the Cold War will need to be complemented by other sensors and surveillance networks today and in the future.[37] First, open sources strongly suggest that few of the arrays installed during the Cold War remain today, either due to inactivity or decay; the arrays emplaced during the Cold War would quickly turn unusable without regular maintenance and electrical current running through the cables between the SOSUS arrays and the terminals ashore.[38] Second, Russia's focus on quality over quantity in its submarine force means that recent and future Russian submarine classes have achieved or will achieve quieting approaching the levels found in U.S. and European submarines; thus, detection ranges will be far shorter than during the Cold War. Furthermore, the likely introduction of Russian AIP technology in the coming years means that future conventional Russian submarines will be able to remain submerged and quiet for much longer than the two or three days that were the norm during the Cold War. In addition, the addition of cruise missiles such as the Kalibr means that Russia's SSNs and SSGNs, and even the smaller Russian conventional submarines, have real and long-range striking power against targets both on land and at sea. The widespread introduction of cruise missiles into the Russian navy in general, and the submarine force in particular, also means that the Northern Fleet will not have to venture as far into the North Atlantic, or south of the GIUK gap, to threaten NATO's reinforcement efforts. It also means that the United States and NATO will have to track and engage Russia's submarines at considerably greater distances than during the Cold War. Also, climate change, which is clearly evident in the far North Atlantic, has made ASW more difficult because of unpredictable changes in water temperatures and salinity levels, which cause changes to sound propagation in the water in ways not fully understood.[39]

New technologies, however, may help NATO and the United States to counter the re-emerging Russian subsurface force in a more effective way. With the emergence of robotics, unmanned systems, big data analytics, and networked platforms, ASW is likely to become a more distributed affair and to move away from its current construct where big and expensive platforms (such as MPAs and submarines) form the bulk of an ASW force. Instead, they could be teamed with smaller and cheaper systems, such as unmanned undersea vehicles that can carry sensors or even be weaponized.[40] Some of the building blocks of the next generation of ASW are currently being fielded in an experimental role.

The Defense Advanced Research Projects Agency (DARPA) is currently developing the Anti-Submarine Warfare Continuous Trail Unmanned Vessel (ACTUV), a surface platform that would be able to stay out at sea for long periods of time and deploy and control sensors over vast distances to detect and track ultra-quiet submarines. Once it leaves the pier it is expected to be able to undertake missions of up to ninety days. Dubbed the Sea Hunter, the 132-foot vessel first put to sea in mid-2016 for sea trials. But more work remains to be done on this concept. The Sea Hunter has to not only be able to detect and track submarines, but also make sense of the gathered data and share it with other units, such as ASW ships and friendly submarines, which may then engage the enemy submarines. In addition, the Sea Hunter will have to make its way through the maritime domain autonomously, all the while being aware of its surroundings and operating safely around, for example, other shipping.[41] Work is also under way on an air-deployed drone that could be air-dropped from an MPA, skim the surface of the ocean, and dive into the water to pursue a suspected submarine.[42] Small unmanned subsurface systems have also been quietly introduced into NATO ASW exercises in the North Atlantic on an experimental basis, but it will be some time before they are full-fledged components of an ASW force.[43] Nonlethal antisubmarine weapons are also being developed to defend ports and other maritime infrastructure against subsurface intrusions or sabotage. While not very helpful in an all-out war, these nonlethal technologies could prove helpful in an age of hybrid warfare in which the sinking of a submarine, or the killing of personnel, would escalate the situation further, and during periods of crises that fall below conventional war.

Unmanned MPAs are also coming into service. These large unmanned MPAs can stay aloft for nearly thirty hours and provide surveillance over vast maritime areas while at high altitude. They can detect and classify radar signals to help build a picture of, for example, enemy air defenses at sea or along a coast. They can also work in tandem with, for example, manned MPAs to share information and data. The U.S. Navy plans to acquire close to seventy unmanned MPAs, and a number of North Atlantic nations are also eyeing it.[44] In the near future, unmanned MPAs could help carry the load of airborne maritime patrol operations, an undertaking that can be tough on human crews due to their duration. So far, however, unmanned vehicles have not offered a solution to the full range of MPA tasks. Large manned MPAs are still the only craft that can not only detect, track, and target ships and submarines at sea, but also attack those targets with antiship missiles or antisubmarine torpedoes. Thus, in the MPA realm, unmanned systems are coming but humans will likely remain aloft looking for submarines for many years to come and attacking them if worse comes to worst, but they will increasingly be supported by unmanned systems above, below, and on the surface of the North Atlantic.[45]

Submarines, both U.S. and allied, will remain key to operating in the North Atlantic to gather intelligence, track Russian naval movements and exercises, and, ultimately, keep the North Atlantic open during a crisis or war in Europe. And while investments are being made in submarine forces on both sides of the Atlantic, it will not be enough to make a real dent in the overall number of submarines, which has dwindled since the end of the Cold War. But this could be offset by pairing submarines with unmanned and autonomous underwater vehicles that could move forward to sense out the undersea space and collect ISR and potentially attack targets.[46] DARPA is hard at work in this field too, with its Mobile Offboard Clandestine Communications and Approach program, which aims to develop small unmanned underwater vehicles that can be deployed ahead of a submarine to detect and target hostile submarines.[47]

Change is also coming to ASW sensors and the equipment used to interpret and exploit the data collected by them. Big data analytics, using software to process and systematize the immense amounts of information and data points being generated by everything from everyday gadgets to advanced sensors, is rapidly changing businesses and everyday life. Using big data analytics can help decision makers and analysts pick out previously unknown patterns of behavior or irregularities, help predict coming events, or better organize an industrial process. There are applications in nearly all walks of life, from the factory floor to the high school classroom. It can even help online dating sites make better matches for those seeking a partner. And global IT companies, including IBM, Oracle, and SAP, are moving in to the business.[48] Big data analytics could also help revolutionize ASW and subsurface operations. High-frequency sonar is normally used to accurately detect and track submarines, but its range is relatively short. Low-frequency sonar can extend much farther in the water but generally does not provide enough precision needed by ASW forces. Big data analytics with enhanced computer power may be able to increase the precision of that low-frequency sonar, meaning that submarine detection ranges can once again be extended out to ranges commonly found during the Cold War.[49] Also, the clutter created by the ambient sounds of the sea, waves, and sea ice grinding together, for example, along with human activity such as fishing and energy extraction, could be scrubbed out by computer analysis, robbing a submarine of the opportunity to hide its own noise signature in the cacophony of the sea.

The combination of unmanned systems, advanced sensors, and long-range weapons may also enable a new form of shore-based ASW. Batteries of ASW weapons based ashore, such as light antisubmarine torpedoes, could be boosted far into the maritime domain by long-range and fast cruise missiles to attack subsurface targets detected and tracked by sensor networks and unmanned platforms.[50] This approach could prove especially viable at choke points in the far North Atlantic.

Taken together, it seems likely that the future of ASW is very much like the developments in other high-end warfare domains: the distribution of sensors and weapons across both manned and unmanned platforms, working together and sharing information in a networked fashion, increasingly working autonomously but in concert, and swarming together at the point of decision.

And while these emerging ASW technologies have applications in all the oceans and seas of the world, the nations concerned with the North Atlantic are especially well placed to take advantage of them. The United States, the United Kingdom, Norway, Germany, Canada, and others all operate sophisticated navies, have advanced defense industries, and have a track record of integrating new technologies into their naval platforms. Indeed, codevelopment of new ASW technologies could be something that the North Atlantic NATO nations could form a consortium around. Still, an effort of this magnitude would carry risks, as the adoption of new technologies and ways of doing business brings uncertainties and untried concepts, something that could prove especially dangerous in an undertaking as serious as ASW against the increasingly sophisticated Northern

Fleet. Technology breakthroughs that would revolutionize ASW, such as the use of lasers, have been heralded before, but they have been found wanting.[51]

Strategic Confusion

While operating in the North Atlantic in the twenty-first century will in many ways be different from the struggles that took place in the region during the twentieth century, the strategic fundamentals will of course remain the same, among them that the North Atlantic remains the maritime connective tissue between North America and Europe that is vital to the effective defense of U.S. friends and allies in Europe. But this fact has become obscured and forgotten over the past two decades. The minds of both policymakers and military leaders have been more focused on the Indian and Pacific Oceans over the past three decades and more recently, after Russia's aggression against Ukraine, on the ground domain in eastern Europe. Furthermore, most naval officers on both sides of the Atlantic entered their careers well after the end of the third battle of the Atlantic during the Cold War. This has opened up intellectual gaps in the understanding of not only the importance of the North Atlantic but also the dynamics of this particular maritime domain, the potential threats, challenges, opportunities, and what actually matters. Indeed, even as concepts such as sea-lanes of communication and maritime choke points have risen in prominence as Washington considers its maritime future in the Pacific and Indian Oceans, many fail to appreciate the North Atlantic as the maritime bridge between North America and Europe.

This lack of understanding about the Atlantic region, save among the smaller nations such as Iceland and Norway that are faced with the North Atlantic on a daily basis, is already making itself felt. A 2017 Washington war game that drew together experts and government officials from across the NATO nations of the North Atlantic on the topic of maritime competition in the North Atlantic revealed the low levels of understanding among both U.S. and European senior decision makers.[52] The strategic conversation about the North Atlantic has been further confused by the emergence of the Arctic as a security policy topic in Washington and elsewhere. The broader Arctic and the North Atlantic are of course related and intertwined, but they are not the same, and Arctic discussions have served to send analysts, military leaders, and politicians off in the wrong direction when thinking about the new Russian challenge in the North Atlantic.

In recent years so-called Arctic security has received increasing attention by decision makers on both sides of the Atlantic, but especially in Washington, London, Ottawa, Copenhagen, and Oslo. Maps and narratives have been created showing the Russian buildup of infrastructure along the northern coastline of Russia, ranging from simple airfields to coast guard stations, and this is taken as evidence of Russia's expansive ambitions in the Arctic. A common recommendation has been to highlight the urgent need for the U.S. Coast Guard to rapidly acquire more icebreakers to operate in the Arctic.[53] Russia's increasing activity deep into its Arctic is, in many ways, beside the point when thinking about Russia's resurgence at sea and its implications for U.S. security and the future of NATO. Much of the infrastructure along Russia's Arctic coastline, far away from the global maritime commons, is simply a case of bringing back infrastructure put in during the Cold War that has crumbled over decades of lack of funds and maintenance.[54] In addition, much of the planned infrastructure is intended for coast guard and legitimate maritime activities in home waters that one could expect of a Russia that hopes the northern sea route will one day become a viable commercial route for global maritime trade. Also, the bases, airfields, and ports in question are too far from the Atlantic and the Pacific to serve as nodes for global or regional power projection. To be sure, there are maritime safety issues in this region that must be attended to. Environmental contamination or a ship in distress would be a serious emergency anywhere in the Arctic and would require a response of a size and complexity that would challenge any Arctic nation. Indeed, much of Russia's vaunted icebreaker fleet is intended for exactly these commercial, rescue at sea, and law enforcement purposes.[55] Russia is indeed installing air defense networks along its Arctic coast, but that is a defense measure against feared nuclear strikes across the North Pole from the United States and does not add to Russia's power projection capability.

But the issue of Arctic security has served to obscure, in terms of high-end naval power and transatlantic security, that the center of gravity has returned to the Kola Peninsula, which today constitutes a high concentration of Russian naval combat and strategic power with the Northern Fleet. It is close to NATO territory and has easy access to the North Atlantic and the maritime domain north of the GIUK gap, where the Northern Fleet can threaten key targets ashore across northern Europe. Additional U.S. icebreakers here would do little to shift the balance or make the United States more able to operate consistently in the region. Unlike in the North American Arctic, the Gulf Stream keeps the waters of the North Atlantic, the Norwegian Sea, and the Barents Sea navigable throughout the year. It is perfectly possible to walk to the shore and take a swim in the Barents Sea in the winter without an ice floe as far as the eye can see. The ice-free zone is actually increasing further because of climate change, which has caused the most rapid warming around the top of the world. The U.S. Coast Guard is certainly deserving of new icebreakers to consistently operate in and around Alaska and the broader Arctic for homeland security purposes, but investments in Arctic capabilities for the Coast Guard has little to no bearing on the problem of the growing power of Russia's Northern Fleet and its access to the North Atlantic from the Kola Peninsula.[56]

Thus, in order to build a twenty-first-century strategy for the fourth battle of the Atlantic—the North Atlantic itself must again be thought of as its own strategic space—related and connected to both the European and North American land mass as well as to the Arctic, of course, but worthy of attention by itself and deserving of crisp understanding of its importance to American and allied national security.

CHAPTER 13

NATO and the United States in the Twenty-First-Century North Atlantic

A Few Principles

BOTH THE UNITED STATES AND NATO must once again prepare to defend the North Atlantic and to deliver deterrence through and from that maritime domain, but must also take into account both the military, political, and technology constants and changes in the North Atlantic region. A renewed allied effort to keep the North Atlantic open in both peace and war should rest on a series of principles, including:

GIVE THE NORTH ATLANTIC ITS DUE.
As a start, the United States should drive the development of a new maritime strategy for NATO. The 2011 maritime strategy served an important purpose of placing maritime issues at the forefront, at least momentarily, at a time when NATO was singularly focused on ground operations far from the sea, but it is now woefully inadequate and focuses on the wrong things. A new NATO maritime strategy should focus on high-end naval challenges and put front

and center the defense of the North Atlantic and the ability to effect maritime reinforcement across it. A good strategy also defines the hard choices that need to be made. Given the increasingly competitive maritime environment, it is time for NATO to bring its focus back to its own backyard and reduce its commitments to maritime expeditionary operations. It will be up to the United States to coordinate and nudge other sets of allies and partners, such as Japan, Australia, South Korea, and Singapore, to help police and have a credible presence in the key maritime domains far from the North Atlantic, such as the Indian Ocean and the Asia-Pacific region. The EU also has a valuable role to play here by keeping a watchful eye on the Mediterranean and the waters off of Africa.

The process of creating a new maritime strategy for NATO is arguably as valuable as the final product itself. It would allow U.S. political and naval leaders to engage with their European counterparts on the future of NATO at sea, and it would serve as an opportunity to highlight the importance of the maritime domain to transatlantic security to audiences who may not spend a lot of time thinking about maritime issues.

THINK INTERNATIONALLY.
History shows that victory in the battles for control of the North Atlantic has gone to the competitor that has been able to marshal the largest amount of resources and the largest number of allies and partners on its side. This will influence the fourth battle of the Atlantic as well and is even more important given the scarcity of resources available. Even as European defense budgets have begun to recover from their long-term decline during the twenty years after the Cold War, the hard fact remains that advanced defense technologies, including naval systems, are increasingly expensive to procure, maintain, and operate. In addition, there is a wide array of air, ground, and maritime capabilities the allies need and are seeking in order to deter Russian aggression, ranging from fifth-generation fighters and long-range air defense to main battle tanks and better communications gear. This means that there is an urgent need to spend available

defense resources in the most efficient way possible. NATO and its members should look to teaming up to build the capabilities and competencies needed to take on the Russian challenge at sea. And some of the building blocks for international cooperation efforts around capabilities for the North Atlantic are already in place. For example, Germany is emerging as a center of excellence for conventional submarine operations within NATO. The German navy has top-notch training centers it is opening up for use by other European allies.[1] Furthermore, German-made submarines can be found in many NATO navies, meaning that they form a NATO standard of sorts for alliance submarines. NATO's members still need to buy their own submarines, of course, but real money could be saved by having a single center for submarine training, simulators, and tactical development. Britain's Royal Navy offers a similar opportunity for allied and partner surface forces with its Flag Officer Sea Training program, including for frigate-sized warships, which are key to ASW in the North Atlantic.

NATO's members could also consider procuring and maintaining high-end but expensive capabilities as a consortium. Future MPAs, for example, could be bought in larger batches, at a lower price, by a group of NATO members who could share infrastructure for maintenance, training, and exercises. Those nations that are too small or resource-constrained to operate MPAs could still contribute to such a consortium by offering forward basing and maintenance facilities so that allied MPAs could quickly deploy between regions.[2] NATO nations around the North Atlantic could also consider investing in commonly held towed array sonar ships, which would add to the sensor networks in the maritime domain. This consortium approach to procurement and maintenance of assets would not be a first for NATO. A group of allies seeking to quickly gain a heavy lift capacity joined together to form a consortium around the procurement and operations of C-17s, with flight hours distributed among the participating nations according to the level of funding they contributed. More recently NATO procured a set of Global Hawk drones to provide airborne ISR support for NATO operations. NATO has, as an alliance, operated Airborne Early Warning and Control Aircraft for decades.[3] And during the Cold War the allies did joint procurement of ASW capabilities, including MPA. A first step toward this approach in the era of renewed competition in the North Atlantic was recently taken in 2017, when Italy, Germany, France, Greece, Spain, and Turkey announced their plan to jointly develop and procure new MPAs.[4]

International cooperation is, of course, rarely easy, and it is especially hard in ASW and submarine operations, which are often considered highly secretive national crown jewels. Indeed, the more a country shares with an ally, even a close one, the greater the risk that vital intelligence will find its way to the opponent, in this case Russia. In the contest for the North Atlantic, this is no idle consideration. The Walker spy ring during the Cold War did significant damage to U.S. efforts to detect and track Soviet submarines in the North Atlantic and elsewhere, and it increased the Soviet understanding of the importance of submarine quieting. Still, this hesitation must be overcome, and the risks of sharing and close cooperation managed, in order to build a true alliance approach to keeping the North Atlantic open in the twenty-first century.

CREATE A NAVAL CLUB OF CLUBS AND DIVIDE THE LABOR.
The United States could also help move along naval coalitions of the willing under the umbrella of NATO. Here America could help encourage regional naval leadership by its European friends and allies by having the United Kingdom take a leadership role in the North Atlantic, while, for example, Germany leads in the Baltic Sea and France or Italy takes on a leading role in the Mediterranean. This would allow the navies of Europe to more directly specialize and focus on the maritime domains of greatest concern to them. In the case of the North Atlantic, a maritime framework could be imagined that would include the United States, the United Kingdom, Norway, Iceland, Denmark, the Netherlands, and, occasionally, Germany and France. This notion of clearer regional responsibility for the maritime domains around Europe is further merited by the many

war games that have been played in Washington and elsewhere since 2014, which have all shown that a first response to a crisis with Russia will likely not be a NATO one, but a regional coalition of the willing led by the United States. NATO would come into play a short while later as the alliance reaches political consensus that is required for collective action. The North Atlantic Treaty, however, does not bar any one nation or group of nations from coming to the aid of an ally without alliance consensus.

As a nation with global interests and commitments and as the leader of NATO, the United States in particular must also consider where to use its increasingly scarce and stretched naval resources in the maritime domains in and around Europe. A sensible division of labor between U.S. maritime forces and the European allies could be a U.S. focus on the North Atlantic north of the GIUK gap, with a secondary effort in the Black Sea, while NATO's northern members in Europe take up the burden in the Baltic Sea. Several factors influence the logic of this arrangement. First, the far North Atlantic is vast, and U.S. nuclear submarines and U.S. ASW systems, such as the P-8, are by far the most capable in this environment. Furthermore, a U.S. focus on the North Atlantic would allow for greater synchronization between the reinforcement shipping coming across the Atlantic and those forces seeking to establish sea control and defend the seaports, airports, and other key nodes in the same area. The Black Sea region, meanwhile, has a number of NATO members with relatively weak and aged navies that could use the help of high-end capabilities of the kind that the U.S. Navy brings. The Baltic Sea, on the other hand, has a number of capable navies already operating there and has far more capacity to handle the small Russian Baltic Sea Fleet. Handling the A2/AD network in Kaliningrad would likely be a job for U.S. air power and cyber capabilities. Besides, the Baltic Sea is an extremely constrained environment, where U.S. naval forces would be at a disadvantage compared to the smaller ships and submarines operated by European navies.[5]

The United States must also consider the best use of its stretched naval resources across the globe. One approach would be to create flexible deterrent forces made up of destroyers and submarines that can both provide long-range precision fire against ground targets and conduct ASW against hostile submarines. The United States could dedicate one of these naval deterrent forces for exercises and forward presence in the North Atlantic.[6]

HARNESS TECHNOLOGY AND PLAN FOR THE SYSTEMS AND NETWORKS OF TOMORROW, NOT ONLY TODAY.
NATO's members must quickly adopt the new technologies becoming available to deter and defeat Russia's new generation of high-quality submarines with standoff weapons. The current efforts to increase the number and presence of MPAs, new submarines, and frigates in the North Atlantic region is a good start, but they must be combined with emerging unmanned systems that will increase the reach and capabilities of the scarce, large platforms and better track the increasingly quiet Russian subsurface force.

But the United States and the alliance should also consider older technologies that may yet serve a cost-effective purpose in the twenty-first century. For example, old-fashioned depth charges and rocket-propelled bombs could play a role in disrupting Russian submarine operations in the North Atlantic. These weapons would be unlikely to score a hard kill against a Russian submarine, but that is beside the point in this particular case. Lessons from the previous battle of the Atlantic show that a submarine that is attacked, and therefore obviously detected, must break contact and leave the area since it has little effective self-defense. This would effectively disrupt the submarine's effort and would do so in a cost-effective way as part of a mix of weapons alongside more sophisticated and new antisubmarine weapons.[7]

DENY THE PURPOSE OF RUSSIA'S NAVAL FORCES IN THE ATLANTIC.
As detailed in this volume, the relatively low number of submarines in Russia's resurgent navy and the evident focus on long-range cruise missiles point to a Russian naval strategy for the North Atlantic that is more aimed at destroying or disrupting the vital port and airport nodes the United States and NATO will need in order to bring in

reinforcements coming across the Atlantic, rather than an all-out campaign to sink allied shipping coming across the maritime domain. Therefore, along with a new focus on detecting, tracking, and, in wartime, engaging Russia's submarines in the North Atlantic, NATO and the United States should consider an element of its strategy to focus on denying the ultimate purpose of Russia's subsurface strategy for the region. This would include a focus on sea- and land-based cruise-missile defense, increasing the resilience of key ports, airfields, and command-and-control nodes, and wider dispersal of forces and bases in northern Europe.

OPERATE UP NORTH.

The new Russian emphasis on long-range strikes from submarines means that Russian submarines will likely not need to operate south of the GIUK gap. Instead, many of the key airports and seaports needed by NATO and the United States are well within range from north of the GIUK gap. This means that during a conflict NATO must seek out and destroy Russian submarines in the far north rather than wait for them to approach the choke point at the GIUK gap. NATO and the United States must therefore exercise and operate to a much larger extent north of the GIUK gap and toward the Barents Sea.[8] The area north of the GIUK gap is one of the world's most challenging in terms of environmental factors and weather. U.S. and NATO member forces must therefore operate and exercise there frequently to rebuild and refine the skills and knowledge to sustain an effective presence there.

Operating north of the GIUK gap and conducting ASW against Russian submarines will put an emphasis on U.S. and allied submarines as the primary ASW combatant, as this area is within reach of Russian air defense and antiship missiles.[9] Such a forward-leaning ASW strategy is, however, not without its complications. It would put NATO's ASW efforts close to the Russian SSBNs based on the Kola Peninsula, and how could Moscow be sure that the ASW campaign was not aimed at Russia's nuclear deterrent? This challenge must be carefully weighed by NATO's current and future leadership.

CONSIDER OPPORTUNITIES TO HOLD THE RUSSIAN BASTION AT RISK.
NATO faces an urgent challenge in the Baltic Sea region, where the Baltic states are small, exposed, and difficult to defend against Russian military aggression. One way to deter Russia from ever seizing one or all of the Baltic states would be to hold key territory or capabilities of high value to Russia at risk. One such region is the Kola Peninsula, with the Northern Fleet and the submarine-based nuclear deterrent. The United States should give serious thought to a naval strategy in the far North Atlantic that in many ways harks back to the maritime strategy of the 1980s. This approach, however, is not without its potential shortcomings. Many NATO allies would be uneasy with this concept, and there is the real risk of pushing Russia into a "use or lose" decision with its submarine-based nuclear deterrent. Still, the concept bears careful consideration and quiet U.S. conversations with key European allies.

PREPARE FOR A MORE CROWDED NORTH ATLANTIC
AND HYBRID WARFARE.
Historically speaking, the North Atlantic has been dominated, in both peace and war, by the nations of the Atlantic region. In the twenty-first century the North Atlantic space will increasingly include actors from as far away as the Pacific, which will bring to bear economic, political, and military influence. This will make the maritime domain in the North Atlantic even busier and will bring new elements of great-power politics into this strategic space. NATO and national decision makers must prepare themselves for increasingly complicated politics in the North Atlantic, which could place strains on the alliance during a crisis.

Furthermore, the fourth battle of the Atlantic, like the other three earlier contests, will hinge on the ability to bring to bear high-end naval power in the maritime domain. The fourth battle, however, will also include hybrid elements, with potential attacks against the vital submarine cables across the North Atlantic, disinformation campaigns, aggression in cyberspace in peacetime with implications in the maritime domain, and disruptions of GPS and communications gear by electronic means.[10] It may also include the use of nonmilitary vessels for military purposes, including coast guard ships, fishing boats, and merchantman.[11] Hybrid warfare has proven to be a tough challenge to deal with for the United States and NATO, as the alliance is, at heart, an institution focused on traditional military challenges. In the case of hybrid warfare in the maritime domain, NATO must focus on building maritime infrastructure resilience, find ways to clearly identify and attribute attacks and provocations, and help educate senior leaders on robust decision making during unclear circumstances. The latter can be accomplished by rigorous political-military wargaming, which would confront decision makers with situations and decision points they may face during a real-world crisis.

NEXT STEPS.

The above outlined principles are not a fully fleshed-out game plan for the United States and NATO to tackle the resurgent Russian navy in the North Atlantic. They are, however, key tenets informed by the lessons learned and the mistakes made during the previous three battles for the Atlantic, combined with developments in technology and geopolitics over the past twenty-five years both in the North Atlantic region and globally. Applying these principles would help the United States and its NATO allies set the right course on what promises to be a long-term struggle, perhaps as long as the last battle of the Atlantic which went on for nearly half a century.

Conclusion

THE TENSION AND FRICTION between the transatlantic community and a Russia under Putin that seeks to tilt the European, and perhaps even the global, security order in its favor is likely to remain for many years to come—at least as long as Putin remains in power in Moscow. The contest is and will remain multidimensional, and Russia will use all elements of national power in pursuit of its goals, including information, economic tools, energy resources, and armed groups with undetermined government linkages. But none of these tools would have much effect if they were not backed up by the implicit, and sometimes explicit, threat of hard power, with armored divisions, long-range bombers, and nuclear submarines. It is therefore now the job of the United States and its NATO allies to restore credible deterrence against armed aggression in Europe. If that fails, the United States and its allies must have in place the plans and the forces to mount an effective defense against a newly nimble and capable Russia. Moscow most probably does not seek an all-out confrontation with America and NATO; the West is, after all, strategically far superior to Russia in terms of military and economic power. But the nightmare scenario remains that an emboldened Putin one day believes that Russia could end NATO and American global leadership for good with a sudden overt armed attack against a NATO country that the alliance would fail to respond to, perhaps at a moment when the United States is busy with a crisis elsewhere in the world, such as on the Korean peninsula or in the Middle East.[1] This would have catastrophic results for the United States, Europe, and the global order. And Russia has momentary time and space advantages in the regions around the Baltic Sea and the Black Sea, and if NATO was stopped or slowed in getting there with real combat power, Moscow might be able to achieve a fait accompli in eastern Europe that would make Washington and NATO hesitate about pressing forward to regain lost NATO territory. And while all reasonable analysis suggests that this would be a high-stakes and risky proposition indeed for Russia, history also instructs that desperate leaders in authoritarian states, worried about their continued hold on power and perhaps poorly advised and divorced from broader perspectives, can make go-for-broke decisions in the face of overwhelming odds. Saddam Hussein, Slobodan Milosevic, and Leopoldo Galtieri, the leader of the junta in Argentina during the Falklands War, come to mind as recent examples. Japan's decision to attack the United States in 1941 to achieve a knockout blow that would deter Washington from responding, even though Japan was clearly strategically inferior to the United States, is another example. Besides, nations and alliances have been known to misunderstand attempts at signaling, and accidents or incidents can be used as an excuse by a nation seeking conflict.

NATO's and America's first reaction to Russia's increased military assertiveness in Europe has been a ground-focused response, with modest forces of armor, infantry, artillery, and their supporting units being rushed to the new frontline states of Poland, the Baltic states, and elsewhere for exercises and to signal U.S. and alliance resolve. This is not only because the front line can be found in eastern Europe, a nearly landlocked part of the world, but it also betrays where NATO's strategic mindset has been for the past thirty years: on the ground or in direct support of ground operations. But as this book goes to press, NATO and the U.S. national security community are reorienting themselves to include the region where the transatlantic alliance really started in 1949: the North Atlantic. But the fourth battle of the Atlantic will be different from the ones that played out during the twentieth century. This one is marked by new geopolitics in the North Atlantic, a Russian navy focused more on quality and long-range weapons than on quantity, European navies needing to catch up in terms of capabilities and numbers while still addressing challenges in places like the Middle East and North Africa, and a U.S. Navy stretched from demands from across the globe as an outcome of an increasingly competitive global environment. The fourth battle of the Atlantic can be won, and peace in Europe can be maintained through deterrence on land, in the air, in cyberspace, and at sea. In the case of the North Atlantic it will require careful use of limited resources, new forms of cooperation among the allies, a careful study of the Russian navy and its strategy and capabilities, a fresh look at ASW in the twenty-first century, and, once again, the alliance taking the maritime domain seriously. Because, as was the case during the first three battles fought in the Atlantic, in the twenty-first century a war over the future of Europe cannot be won in the North Atlantic, but it certainly can be lost there.

Notes

Introduction

1. Bill Gardner, "Did Russian Submarine Nearly Drag Scottish Fishing Trawler to Watery Grave?," *Guardian*, March 21, 2015.
2. "Finland Drops Depth Charges in 'Submarine' Alert," *BBC News*, April 28, 2015.
3. "Russian Military Jet Nearly Collides With Passenger Plane—Again," *Deutsche Welle*, December 13, 2014.

Chapter 1. An Introduction to the North Atlantic

1. The section on the historical role of the North Atlantic was in large part derived from James Stavridis, *Sea Power: The History and Geopolitics of the World's Oceans* (New York: Penguin Books, 2017), 43–86. For further reading on the Atlantic and its importance to the development of the modern world, see Simon Winchester, *Atlantic: Great Sea Battles, Heroic Discoveries, Titanic Storms, and a Vast Ocean of a Million Stories* (New York: HarperCollins, 2011).

Chapter 2. Submarines

1. See, for example, Sherry Sontag and Christopher Drew, *Blind Man's Bluff: The Untold Story of American Submarine Espionage* (New York: Public Affairs, 2016).
2. For a more detailed discussion about the capabilities and vulnerabilities of submarines, see John Stillion and Bryan Clark, "What It Takes to Win: Succeeding in 21st Century Battle Network Competitions," Center for Strategic and Budgetary Assessments, 2015, 5–7.
3. See John Keegan, *The Price of Admiralty* (London: Hutchison, 1988).
4. For an excellent overview of military operations during the Falklands War, and especially naval operations and the role of submarines, see Max Hastings and Simon Jenkins, *The Battle for the Falklands* (New York: W. W. Norton & Company, 1984).

Chapter 3. World War I

1. Jan Breemer, "Defeating the U-Boat: Inventing Anti-Submarine Warfare," Naval War College, 2010, 16.
2. P. W. Singer, *Wired for War: The Robotics Revolution and Conflict in the 21st Century* (New York: Penguin Press), 158.
3. Erik Larson, *Dead Wake: The Last Crossing of the Lusitania* (New York: Crown Publishing, 2015), 146.
4. Breemer, *Defeating the U-Boat*, 35.
5. Breemer, 47.
6. See Larson, *Dead Wake.*
7. Breemer, *Defeating the U-Boat*, 40–41.
8. Owen Cote, "The Third Battle: Innovation in the US Navy's Silent Cold War Struggle with Soviet Submarines," *Naval War College Newport Papers* 16 (2003): 6.
9. Breemer, *Defeating the U-Boat*, 57.
10. Breemer, 47–51.
11. Charles Maynard, *The Murmansk Venture* (New York: Arno Press, 1971), 8–9; and Max Boot, *The Savage Wars of Peace: Small Wars and the Rise of American Power* (New York: Basic Books, 2014), 207–11.
12. Bryant Ranft and Geoffrey Till, *The Sea in Soviet Strategy* (Annapolis, Md.: Naval Institute Press, 1983), 171.

Chapter 4. World War II

1. Owen Cote, "The Third Battle: Innovation in the US Navy's Silent Cold War Struggle with Soviet Submarines," *Naval War College Newport Papers* 16 (2003): 9.
2. Cote, "The Third Battle," 8–9.
3. Rick Atkison, *An Army at Dawn: The War in North Africa, 1942–1943* (New York: Holt, 2007), 22.
4. Winston Churchill, *The Second World War: Closing the Ring* (Boston: Mariner Books, 1986).
5. Cote, "The Third Battle," 12.
6. Norman Polmar, *Guide to the Soviet Navy* (Annapolis, Md.: Naval Institute Press, 1986), 2.
7. Earl Ziemke, "The German Northern Theater of Operations 1940–1945," Department of the Army, 1959, 2–13.
8. Hans W. Weigert, "Iceland, Greenland and the United States," *Foreign Affairs* 23, no. 1 (October 1944): 112–13.
9. Charles Emmerson, *The Future History of the Arctic* (New York: Public Affairs, 2010), 106–7.
10. Cote, "The Third Battle," 78.
11. See Arthur Herman, *Freedom's Forge: How American Business Produced Victory in World War II* (New York: Random House, 2013).

12. John Stillion and Bryan Clark, "What It Takes to Win: Succeeding in 21st Century Battle Network Competitions," Center for Strategic and Budgetary Assessments, 2015, 12.

Chapter 5. The Cold War

1. Norman Polmar, *Guide to the Soviet Navy* (Annapolis, Md.: Naval Institute Press, 1986), 14.

2. Norman Polmar and Edward Whitman, *Hunters and Killers,* vol. 2: *Anti-Submarine Warfare Since 1943* (Annapolis, Md.: Naval Institute Press, 2016), 78.

3. Bryant Ranft and Geoffrey Till, *The Sea in Soviet Strategy* (Annapolis, Md.: Naval Institute Press, 1983), 200.

4. Ranft and Till, *The Sea in Soviet Strategy,* 200.

5. Ranft and Till, 119.

6. Owen Cote, "The Third Battle: Innovation in the US Navy's Silent Cold War Struggle with Soviet Submarines," *Naval War College Newport Papers* 16 (2003): 38–39.

7. See, for example, Jerry Hendrix, "Retreat from Range: The Rise and Fall of Carrier Aviation," Center for a New American Security, 2015.

8. Ranft and Till, *The Sea in Soviet Strategy,* 125–28.

9. Polmar, *Guide to the Soviet Navy,* 26–27.

10. See Sam Tangredi, *Anti-Access Warfare : Countering A2/AD Strategies* (Annapolis, Md.: Naval Institute Press, 2013), 151.

11. For a lengthier discussion about the role of submarines and lessons learned from the Falklands campaign, see Anthony Wells, *A Tale of Two Navies: Geopolitics, Technology, and Strategy in the United States Navy and the Royal Navy, 1960–2015* (Annapolis, Md.: Naval Institute Press, 2017), 140–57; and Max Hastings and Simon Jenkins, *The Battle for the Falklands* (New York: W. W. Norton & Company, 1984).

12. Barry D. Watts, "The Maturing Revolution in Military Affairs," Center for Strategic and Budgetary Assessments, 2011, 22–23.

13. Ranft and Till, *The Sea in Soviet Strategy,* 165.

14. Polmar, *Guide to the Soviet Navy,* 40.

15. Gustav Petursson, "Iceland Security," in *Security and Sovereignty in the North Atlantic,* ed. Lassi Heininen (Basingstoke, U.K.: Palgrave MacMillan, 2014), 30.

16. Ranft and Till, *The Sea in Soviet Strategy,* 175.

17. Ranft and Till, 140–41.

18. Tomas Ries, "Soviet Military Strategy and Northern Waters," in *The Soviet Union and Northern Waters,* ed. Clive Archer (London: Royal Institute of International Affairs, 1988), 90–92.

19. Polmar, *Guide to the Soviet Navy,* 41.

20. See Mikhail Morukov, "The White Sea-Baltic Canal," in *The Economics of Forced Labor: The Soviet Gulag,* ed. Paul Gregory and V. Lazarev (Stanford, Calif.: Stanford University Press, 2003), 151–62.

21. See Gennady Luzin, Michael Pretes, and Vladimir Vasiliev, "The Kola Peninsula: Geography, History and Resources," *Arctic* 47, no. 1 (March 1994): 1–15; and Ranft and Till, *The Sea in Soviet Strategy*, 132.

22. Ranft and Till, *The Sea in Soviet Strategy*, 132.

23. U.S. Office of Technology Assessment, "Nuclear Wastes in the Arctic: An Analysis of Arctic and Other Regional Impacts from Soviet Nuclear Contamination," (Washington, D.C.: Government Printing Office, 1995), 120–22.

24. Polmar, *Guide to the Soviet Navy*, 468.

25. Oleg Bukharin and Joshua Handler, "Russian Nuclear-Powered Submarine Decommissioning," *Science and Global Security* 5 (1995): 249–52.

26. Polmar, *Guide to the Soviet Navy*, 469.

27. David Winkler, *Cold War at Sea: High-Seas Confrontations between the United States and the Soviet Union* (Annapolis, Md.: Naval Institute Press, 2000), 119.

28. See Dean Allard, "Strategic Views of the US Navy and NATO on the Northern Flank 1917–1991," *Northern Mariner* 11, no. 1 (January 2001): 11–24.

29. Polmar and Whitman, *Hunters and Killers,* 2:126.

30. Cote, "The Third Battle," 41–42.

31. See Dawn Maskell, "The Navy's Best-Kept Secret: Is IUSS Becoming a Lost Art?," U.S. Marine Corps Command and Staff College, 2001.

32. Cote, "The Third Battle," 32.

33. William Broad, "Scientists Fight Navy Plan to Shut Far-Flung Undersea Spy System," *New York Times*, July 12, 1994.

34. Anthony Wells, *A Tale of Two Navies: Geopolitics, Technology, and Strategy in the United States Navy and the Royal Navy, 1960–2015* (Annapolis, Md.: Naval Institute Press, 2017), 24.

35. Malcolm Chalmers, "The UK and the North Atlantic After Brexit," in *NATO and the North Atlantic: Revitalising Collective Defense,* ed. John Andreas Olsen (London: Royal United Services Institute, 2017), 32–33.

36. Clive Archer, "Norwegian Sea and Northern Waters: British Views Since the 1970s," in *Soviet Sea Power in Northern Waters*, ed. John Skogan and Arne Brundtland (New York: St. Martin's Press, 1990), 77.

37. F. U. Kupferschmidt, "A German View," in *Britain and NATO's Northern Flank,* ed. Geoffrey Till (New York: St. Martin's Press, 1988), 103–9.

38. Jay Wagner, "The West German Response to Soviet Naval Activity in the North," in *Soviet Seapower in Northern Waters,* ed. John Kristen Skogan and Arne Olav Brundtland (New York: St. Martin's Press 1990), 130–47.

39. Rolf Tamnes, "The Significance of the North Atlantic and the Norwegian Contribution," in *NATO and the North Atlantic: Revitalising Collective Defense,* ed. John Andreas Olsen (London: Royal United Services Institute, 2017), 16.

40. Johan Jorgen Holst, "Strategic Developments in the North Atlantic and the Norwegian Sea: Challenges to Norway," in *Soviet Sea Power in Northern Waters*, ed. John Skogan and Arne Brundtland (New York: St. Martin's Press, 1990), 184.

41. Beinta Jakupsstovu, "The Faroe Islands' Security Policy in a Process of Devolution," *Iceland Review of Politics & Administration* 8, no. 2 (2012): 417–18.

42. David Fouquet, "U.S., NATO Build Up Bases in Azores, Madeira," *New York Times*, April 30, 1987.

43. Robert Harkavy, "Bases Abroad: The Global Foreign Military Presence," Stockholm International Peace Research Institute, 1989, 83–86.

44. William Perkins, "Alliance Airborne Anti-Submarine Warfare," Joint Air Power Competency Center, 2016, 13–14.

45. Tamnes, "The Significance of the North Atlantic and the Norwegian Contribution," 16–17.

46. Marco Borst, "Orions of the Netherlands," *Airborne Log* (Fall 1994): 12–15.

47. See NATO Undersea Research Center, "Taking the Future to Sea: 50th Anniversary 1959–2009," La Spezia, Italy, 2009.

48. Alan Draper, *European Defense Equipment Collaboration: Britain's Involvement, 1957–87* (Berlin: Springer, 1990), 15–21.

49. Aviation Safety Network, "Accident Description: Breguet SP-13A Atlantic Thursday 15 January 1981," http://aviation-safety.net/database/record.php?id=19810115-1.

50. American University, "TED Case Studies: Cod War," May 1997, http://www1.american.edu/ted/icefish.htm.

51. See Valur Ingimundarson, "Fighting the Cod Wars in the Cold War: Iceland's Challenge to the Western Alliance in the 1970s," *RUSI Journal* 148, no. 3: 88–94.

52. Charles Emmerson, *The Future History of the Arctic* (New York: Public Affairs, 2010), 291–93.

53. Simon Duke, "United States Forces and Military Installations in Europe," Stockholm International Peace Research Institute, 1989, 181–91.

54. James Markham, "Iceland's Elves Are Enlisted in Anti-NATO Effort," *New York Times*, March 30, 1982.

55. Christian Atland and Torbjorn Pedersen, "Cold War Legacies in Russia's Svalbard Policy," in *Environmental and Human Security in the Arctic*, ed. Gunhild Hoogensen Gjorv (London: Routledge, 2013), 23.

56. David Winkler, "Tuesday's Buzzing Had Deadly Precedent," Naval Historical Foundation, April 15, 2016.

57. Havard Klevberg, *Request Tango: 333 skvadron pa ubatsjakt—maritime luftoperasjoner i norsk sikkerhetspolitikk* (Oslo: Universitetsforlaget, 2012), 406–8.

58. Drew Middleton, "NATO Naval Exercise Off Norway under Intensive Soviet Surveillance," *New York Times*, September 28, 1979.

59. See Winkler, *Cold War at Sea*.

60. Elizabeth Pond, "Svalbard Arctic Outpost at Strategic Crossroads," *Christian Science Monitor*, September 9, 1980; and Jeffrey Barlow, "NATO's Northern Flank: The Growing Soviet Threat," Heritage Foundation, May 1979, 1–2.

61. "Soviet Cruise Missile Said to Stray across Norway and into Finland," *New York Times*, January 3, 1985.

62. See, for example, John Hattendorf, "The Evolution of the US Navy's Maritime Strategy, 1977–1986," *Naval War College Newport Papers* 19 (2004); and Peter Swartz, "Understanding an Adversary's Strategic and Operational Calculus: A Late Cold War Case Study with 21st Century Applicability," Center for Naval Analyses, 2013.

63. See John Lehman, *Oceans Ventured: Winning the Cold War at Sea* (New York: W. W. Norton & Company, 2018).

64. See Richard Hooker, "NATO's Northern Flank: A Critique of the Maritime Strategy," *Parameters* (June 1989): 24–37.

65. See Allard, "Strategic Views of the US Navy and NATO," 11–24.

Chapter 7. Rust and Nuclear Waste

1. Hal Brands, "Dealing with Allies in Decline," Center for Strategic and Budgetary Assessments, 2017, 6.

2. Charles Krupnick, *Decommissioned Russian Nuclear Submarines and International Cooperation* (Jefferson, N.C.: McFarland & Company, 2001), 55.

3. Marlene Laruelle, *Russia's Arctic Strategies and the Future of the Far North* (Abingdon, U.K.: Routledge, 2014), 120.

4. Susanne Kopte, "Nuclear Submarine Decommissioning and Related Problems," Bonn International Center for Conversion, 1997, 21.

5. Roald Gjeltsen, "The Role of Naval Forces in Northern Waters at the Beginning of a New Century," in *Navies in Northern Waters 1721–2000*, ed. Rolf Hobson and Tom Kristiansen (Milton Park, U.K.: Taylor and Francis, 2004), 281.

6. Krupnick, *Decommissioned Russian Nuclear Submarines*, 72.

7. Igor Kudrik, "Shoot-Out on Nuclear Powered Submarine," Bellona Foundation, September 11, 1998.

8. See Timothy Heleniak, "Boom and Bust: Population Change in Russia's Arctic Cities," policy note presented at George Washington University, May 30–31, 2013.

9. Ingemar Dorfer, "Kola Has Lost Significance," U.S. Naval Institute *Proceedings* 128 (March 2002).

10. Laruelle, *Russia's Arctic Strategies and the Future of the Far North*, 113.

11. Oleg Bukharin and Joshua Handler, "Russian Nuclear-Powered Submarine Decommissioning," *Science and Global Security* 5 (1995): 254.

12. Bryant Ranft and Geoffrey Till, *The Sea in Soviet Strategy* (Annapolis, Md.: Naval Institute Press, 1989), 150.

13. Thomas Nielsen, "Sevmash Loses Yet Another Prestige Contract," *Barents Observer*, August 12, 2008.

14. Kopte, "Nuclear Submarine Decommissioning and Related Problems," 12.

15. Kopte, 1.

16. Krupnick, *Decommissioned Russian Nuclear Submarines*, 38.

17. Kopte, "Nuclear Submarine Decommissioning and Related Problems," 1.

18. Charles Digges, "Russia Announces Enormous Finds of Radioactive Waste and Nuclear Reactors in Arctic Sea," Bellona Foundation, August 28, 2012.

19. Thomas Nielsen, "Last Cold War Submarine Ready for Scrapping," *Barents Observer*, July 12, 2012.

20. Robert Beckhusen, "Russia Is Finally Slicing Up Its Abandoned, Contaminated Submarines," *War Is Boring*, October 17, 2013.

21. Krupnick, *Decommissioned Russian Nuclear Submarines*, 31.

22. Havard Klevberg, *Request Tango: 333 skvadron pa ubatsjakt—maritime luftoperasjoner i norsk sikkerhetspolitikk* (Oslo: Universitetsforlaget, 2012), 374.

23. The section on the events surrounding the *Kursk* submarine disaster is derived from multiple sources, including Robert More, *A Time to Die* (New York: Random House, 2003); Clyde Burleson, *Kursk Down: The Shocking True Story of the Sinking of a Russian Nuclear Submarine* (New York: Warner Books, 2002); and Peter Truscott, *Kursk* (New York: Simon & Schuster, 2002).

24. Fiona Hill and Clifford Gaddy, *Mr. Putin: Operative in the Kremlin* (Washington, D.C.: Brookings Institution, 2013), 180.

25. See, for example, NATO Maritime Command, "Russian Federation Navy Fully Integrated in NATO Submarine Rescue Exercise Bold Monarch," news release, June 7, 2011.

Chapter 8. High North, Low Tension

1. Trude Pettersen, "Closer Military Cooperation between Norway and Russia," *Barents Observer*, February 13, 2013.

2. Andreas Osthagen, "High North, Low Politics—Maritime Cooperation with Russia in the Arctic," *Arctic Review on Law and Politics* 7, no. 1 (2016): 85–87.

3. Marco Borst, "Ilyushin-38 MAY—The Russian Orion," *Airborne Log* (Summer 1996): 8–9.

4. Charles Krupnick, *Decommissioned Russian Nuclear Submarines and International Cooperation* (Jefferson, N.C.: McFarland & Company, 2001), 138.

5. John Cushman, "US Periscopes Still Follow Soviet Fleet," *New York Times*, February 23, 1992.

6. Gustav Petursson, "Iceland Security," in *Security and Sovereignty in the North Atlantic*, ed. Lassi Heininen (Basingstoke, U.K.: Palgrave Macmillan, 2014), 38–39.

7. Fred Hiatt, "U.S., Russian Navies to Conduct First Joint Exercise," *Washington Post*, July 1, 1992.

8. Lily Daniels, "US, Russian Navies Complete Northern Eagle 2003," *US Naval Forces Europe*, October 6, 2004.

9. Trude Pettersen, "Exercise Northern Eagle Has Started," *Barents Observer*, August 20, 2016.

10. John Lehman, *Oceans Ventured: Winning the Cold War at Sea* (New York: W. W. Norton & Company, 2018), 271.

11. Andrew Zask, "New International Submarine Rescue Coordination Center Opens," US Atlantic Fleet Public Affairs, September 29, 2004.

12. Gunhild Hoogensen Gjorv, ed., *Environmental and Human Security in the Arctic* (London: Routledge, 2013).

13. Abbie Tingstad et al., "Will the Breakdown in US-Russia Cooperation Reach the Arctic?," RAND Corporation, October 12, 2016.

14. Trude Pettersen, "Snow Crabs Have Found Niche in Barents Sea Ecosystem," *Barents Observer*, March 23, 2014.

15. Atle Staalesen, "Snow Crabs Raise Conflict Potential around Svalbard," *Barents Observer*, January 23, 2017.

16. Roald Gjeltsen, "The Role of Naval Forces in Northern Waters at the Beginning of a New Century," in *Navies in Northern Waters 1721–2000*, ed. Rolf Hobson and Tom Kristiansen (Milton Park, U.K.: Taylor and Francis, 2004), 298.

Chapter 9. Pirates, Terrorists, and the Hindu Kush

1. Gary Weir and Sandra J. Doyle, eds., *You Cannot Surge Trust: Combined Naval Operations of the Royal Australian Navy, Canadian Navy, Royal Navy, and United States Navy, 1991–2003* (Washington, D.C.: Naval History and Heritage Command, 2015), 63–65; and Stacy Poe, "Rules of Engagement: Complexities of Coalition Interactions in Operations Other Than War," Defense Technical Information Center, 1995, 13–14.

2. See, for example, David Rieff, *Slaughterhouse: Bosnia and the Failure of the West* (New York: Touchstone, 1996).

3. Javier Solana, "NATO: Its 50th Anniversary—The Washington Summit—The Next Century," North Atlantic Treaty Organization, Brussels, Belgium, January 25, 1999.

4. Bradley Graham, "Abizaid Credited with Popularizing the Term 'Long War,'" *Washington Post*, February 3, 2006.

5. Gerry Gilmore, "Rumsfeld: Transformation Moving NATO into 21st Century," Armed Forces Press Service, October 14, 2004, http://archive.defense.gov/news/newsarticle.aspx?id=25073.

6. See Robert Jordan, *Alliance Strategy and Navies: The Evolution and Scope of NATO's Maritime Dimension* (New York: St. Martin's Press, 1990); and Geoffrey Till, "Holding the Bridge in Troubled Times: The Cold War and the Navies of Europe," *Journal of Strategic Studies* 28, no. 2 (2005): 316.

7. See, for example, Ivo Daalder and James Goldgeier, "Global NATO," *Foreign Affairs*, September/October 2006; and Corry Kucik, "NATO's Future: It's Far from the North Atlantic," U.S. Naval Institute *Proceedings* 140 (September 2014).

8. See "Active Engagement, Modern Defence: Strategic Concept for the Defence and Security of the Members of the North Atlantic Treaty Organisation," adoped by the heads of State and Government in Lisbon, November 19–20, 2010, https://www .nato.int/nato_static_fl2014/assets/pdf/pdf_publications/20120214_strategic -concept-2010-eng.pdf..

9. David Axe, "Pirate-Fighers, Inc: How Mercenaries Became Ships' Best Defense," *Wired*, August 23, 2011.

10. See Mark Barrett, et al., "Assured Access to the Global Commons," Allied Command Transformation, 2010.

11. North Atlantic Treaty Organization, "Alliance Maritime Strategy," Brussels, Belgium, March 18, 2011, http://www.nato.int/cps/en/natohq/official_texts_75615 .htm.

12. See Brooke Smith-Windsor, "NATO's Maritime Strategy and the Libya Crisis as Seen from the Sea," NATO Defense College, Rome, 2013.

13. See "Active Engagement, Modern Defence."

14. See "European Union Maritime Security Strategy," Brussels, Council of the European Union, June 24, 2014.

15. See "A Cooperative Maritime Strategy for 21st Century Seapower," U.S. Department of Defense, October 2007.

16. U.S. European Command, "Russian Destroyer RFS Natoychiviy (DD 610) Is Currently Underway Off the Coast of Ventspils, Latvia, on June 6, 2005, Participating in Baltic Operations (BALTOPS) 2005," June 6, 2005, http://www.eucom.mil /media-library/photo/18571/eucom-photo.

17. Kathleen Hicks et al., "Undersea Warfare in Northern Europe," Center for Strategic and International Studies, 2016, 23.

18. Information Dissemination, "The Holland-Class OPVs Will Need to Make a Change," March 12, 2010, http://www.informationdissemination.net/2010/03 /holland-class-opvs-will-need-change.html.

19. F. Stephen Larrabee et al., "NATO and the Challenges of Austerity," RAND Corporation, 2012, 11.

20. Larrabee et al., 31–32.

21. Larrabee et al., 8, 28, 37, 46, 53.

22. Julianne Smith and Jerry Hendrix, "Forgotten Waters: Minding the GIUK Gap," Center for New American Security, 2017, 8.

23. See Bryan McGrath, "NATO at Sea: Trends in Allied Seapower," American Enterprise Institute, 2013.

24. See Her Majesty's Government, "Securing Britain in an Age of Uncertainty: The Strategic Defence and Security Review," Cabinet Office, London, October 2010.

25. William Perkins, "Alliance Airborne Anti-Submarine Warfare," Joint Air Power Competency Center, June 2016, 51.

26. Norman Polmar and Edward Whitman, *Hunters and Killers,* vol.2: *Anti-Submarine Warfare from 1943* (Annapolis, Md.: Naval Institute Press, 2016), 198.

27. Hicks et al., "Undersea Warfare in Northern Europe," 21.

28. "Norway's Sub-Snub Impacts Negatively on Nordic on Nordic Defense Cooperation," *Defense News,* April 28, 2016.

29. Larrabee et al., "NATO and the Challenges of Austerity," 31–32.

30. Michael Whitby, "Odd Jobs: Canada's Use of Submarines on Fisheries Patrols 1993–1995. Part I," *Submarine Review* (September 2017): 65–71.

31. Magnus Nordenman, "The Incredible, Shrinking Modern Military," *Atlantic Monthly,* November 12, 2012.

32. "Germany's Lack of Military Readiness 'Dramatic' Says Bundeswehr Commissioner," *Deutsche Welle,* February 20, 2018.

33. Tom Clancy, *Carrier: A Guided Tour of an Aircraft Carrier* (New York: Berkley Books, 1999), 178–82; and Polmar and Whitman, *Hunters and Killers,* vol 2: 196.

34. Daniel Dolan, "When ASW Didn't Matter," U.S. Naval Institute *Proceedings* (August 2017): 60–63.

35. Nicholas Woodworth, "Rebuild Air ASW," U.S. Naval Institute *Proceedings* (October 2017): 33.

36. "Silent Threat," *NBC-4* news broadcast, October 26, 2006.

37. "Sweden Submarine Capabilities," *NTI,* July 23, 2013, http://www.nti.org/analysis/articles/sweden-submarine-capabilities/.

38. Polmar and Whitman, *Hunters and Killers,* 2: 202.

39. United States House of Representatives, Subcommittee on Defense, Committee on Appropriations, *Statement of Admiral Jonathan Greenert on FY 2016 Department of the Navy Posture,* February 26, 2015, 17.

40. Karen Deyoung and Greg Jaffe, "NATO Runs Short of Some Munitions in Libya," *Washington Post,* April 15, 2011.

41. Gregory Pedlow, "The Evolution of NATO's Command Structure, 1951–2009," NATO, 2010, 14.

42. Elizabeth Braw, "The Secret Norwegian Submarine Base Being Rented by the Russians," *Newsweek,* March 19, 2015.

43. Adam Herbert, "Presence, Not Permanence," *Air Force* (August 2006): 34–39.

44. Gustav Petursson, "Iceland Security," in *Security and Sovereignty in the North Atlantic,* ed. Lassi Heininen (Basingstoke, U.K.: Palgrave Macmillan, 2014), 30–31.

45. Paul Almes, "China's Atlantic Stop-Over Worries Washington," *Politico,* September 29, 2016.

46. Seth Cropsey, "Restore the US Sixth Fleet," *National Review,* November 2, 2015.

47. Luis Rodriguez and Sergiy Glebov, *Military Bases: Historical Perspectives, Contemporary Challenges* (Amsterdam: IOS Press, 2009), 136–38.

48. Polmar and Whitman, *Hunters and Killers,* 2: 127.

49. William Broad, "Scientists Fight Navy Plan to Shut Far-Flung Undersea Spy System," *New York Times,* June 12, 1994.

Chapter 10. The Return of Competition in the North Atlantic

1. Martin Murphy, Gary Schaub, and Frank Hoffman, "Hybrid Maritime Warfare and the Baltic Sea Region," Center for Military Studies, University of Copenhagen, 2016, 5.

2. "French Navy Spots Nuclear-Armed Submarine off Coast," *Reuters*, March 10, 2016.

3. William Perkins, "Alliance Airborne Anti-Submarine Warfare," Joint Air Power Competency Center, June 2016, 22–23.

4. Iceland Coast Guard, Reykjavik, Iceland, briefing to the author, November 15, 2016.

5. Bradley Peniston, "A Key NATO Ally Looks Nervously at Putin—and Trump," *Defense One*, January 23, 2017.

6. Marlene Laruelle, *Russia's Arctic Strategies and the Future of the Far North* (Abingdon, U.K.: Routledge, 2014), 126.

7. Laruelle, 124–25.

8. Sean O'Connor, "Russian Carrier Jets Flying from Syria, Not Kuznetsov," *Jane's Defence Weekly*, November 28, 2016.

9. "New Submarine Rostov-on-don Carried Out Missile Launches in the Barents Sea," press release, Russian Ministry of Defense, October 2, 2015.

10. Christopher Cavas, "Russian Submarine Hits Targets in Syria," *Defense News*, December 8, 2015.

11. Stuart Williams, "Despite Tensions, Russia's 'Syria Express' Sails by Turkey," *Digital Journal*, January 5, 2016.

12. "Veliky Novgorod and Kolpino Submarines Fired the Kalibr Cruise Missiles from Submerged Position against the ISIS Critical Objects in Syria," Russian Ministry of Defense, September 17, 2017.

13. Perkins, "Alliance Airborne Anti-Submarine Warfare," 23.

14. Hans Kristensen, "Kalibr: Savior of INF Treaty?," Federation of American Scientists, December 14, 2015.

15. Jeffrey Lewis, "Sokov on Russian Cruise Missiles," *Arms Control Wonk*, August 25, 2015.

16. Christopher Cavas, "Is Caspian Sea Fleet a Game-Changer?" *Defense News*, October 11, 2015.

17. Defense Intelligence Ballistic Missile Analysis Committee, "Ballistic and Cruise Missile Threat 2017," U.S. Department of Defense, 35.

18. Carlo Kopp, "Defeating Cruise Missiles," *Australian Aviation*, October 2004, 3–4.

19. Roger N. McDermott and Tor Bukkvoll, *Russia in the Precision-Strike Regime – Military Theory, Procurement and Operational Impact* (Oslo: Norwegian Defense Research Establishment, 2017), 14.

20. McDermott and Bukkvoll, 32.

21. Lee Willett, "Game Changer: Russian Sub-Launched Cruise Missiles Bring Strategic Effect," *Jane's International Defence Review*, April 27, 2017, 5.

22. Willett, 5.

23. McDermott and Bukkvoll, *Russia in the Precision-Strike Regime*, 19.

24. Rolf Tamnes, "The Significance of the North Atlantic and the Norwegian Contribution," in *NATO and the North Atlantic: Revitalising Collective Defense,* ed. John Andreas Olsen (London: Royal United Services Institute, 2017), 25.

25. Willett, "Game Changer," 7.

26. "Russia Military Power: Building a Military to Support Great Power Aspirations," U.S. Defense Intelligence Agency, 2017, 70.

27. Barry D. Watts, "The Maturing Revolution in Military Affairs," Center for Strategic and Budgetary Assessments, 2011, 14.

28. Mark Gunzinger et al., "Force Planning for the Era of Great Power Competition," Center for Strategic and Budgetary Assessments, 2017, 84.

29. "Russian Military Power," 22.

30. Watts, "The Maturing Revolution in Military Affairs," 1.

31. Bryant Ranft and Geoffrey Till, *The Sea in Soviet Strategy* (Annapolis, Md.: Naval Institute Press, 1989), 210–11.

32. See John Altmann, "Russian A2/AD in the Eastern Mediterranean: A Growing Risk," *Naval War College Review* 69, no. 1 (Winter 2016): 72–84.

33. Department of Defense, Department of Defense press briefing with General Breedlove in the Pentagon Briefing Room, March 1, 2016, https://www.defense .gov/News/Transcripts/Transcript-View/Article/683817/department-of-defense -press-briefing-by-gen-breedlove-in-the-pentagon-briefing.

34. Thomas Frear, "Anatomy of a Russian Exercise," European Leadership Network, August 12, 2015.

35. Ryan Faith, "Russia's Massive Military Exercise in the Arctic Is Utterly Baffling," *Vice News*, March 20, 2015.

36. Julian Ropke, "Putin's Zapad 2017 Simulated a War against NATO," *Bild*, December 19, 2017.

37. Kyle Mizokami, "Russian Is Bringing Back Blackjack, the Last Soviet Bomber," *Popular Mechanics*, June 12, 2017; and Valery Konyshev and Alexander Sergunin, "Russian Military Strategies in the High North," in *Security and Sovereignty in the North Atlantic,* ed. Lassi Heininen (Basingstoke, U.K.: Palgrave Macmillan, 2014), 91.

38. Bryan Clark, Center for Strategic and Budgetary Assessments, interview with the author, Washington, D.C., April 24, 2017.

39. Andrew Metrick and Kathleen Hicks, "Contested Seas: Maritime Domain Awareness in Northern Europe," Center for Strategic and International Studies, 10.

40. For more on potential cruise missile strategies, see David Nicholls, "Cruise Missiles and Modern War: Strategic and Technological Implications," Center for Strategy and Technology, Air War College, Occasional Paper No. 13, 13–20.

41. Steve Wills, "A New Gap in the High North and Forward Defense Against Russian Naval Power," Center for International Maritime Security, July 17, 2018.

42. McDermott and Bukkvoll, *Russia in the Precision-Strike Regime*, 33.

43. David Sanger and Eric Schmitt, "Russian Ships Near Data Cables Too Close for US Comfort," *New York Times*, October 25, 2015.

44. David Winkler, *Cold War at Sea: High-Seas Confrontations between the United States and the Soviet Union* (Annapolis, Md.: Naval Institute Press, 2000), 38; and Norman Polmar, *Guide to the Soviet Navy* (Annapolis, Md.: Naval Institute Press, 1986), 22.

45. "Trans-Atlantic Bandwidth: Then and Now," *War on the Rocks*, October 30, 2015.

46. Tara Davenport, "Submarine Cables, Cybersecurity and International Law: An Intersectional Analysis," *Catholic University Journal of Law and Technology* 24, no. 1 (December 2015): 61.

47. Alison Russell, "Strategic Anti-Access/Area-Denial in Cyberspace," conference paper for 7th International Conference on Cyber Conflict, NATO Cooperative Cyber Defence Centre of Excellence, 2015, 159.

48. Jeremy Scahill, *The Assassination Complex: Inside the Government's Secret Drone Warfare Program* (New York: Simon & Schuster, 2016), 73.

49. Russell, "Strategic Anti-Access/Area-Denial in Cyberspace," 158.

50. Ingrid Burrington, "What's Important About Underwater Internet Cables," *Atlantic*, November 9, 2015.

51. Jonathan Reed Winkler, "Silencing the Enemy: Cable-Cutting during the Spanish-American War," *War on the Rocks*, November 5, 2015.

52. Jonathan Reed Winkler, *Nexus: Strategic Communications and American Security in World War I* (Cambridge, Mass.: Harvard University Press, 2008), 5.

53. See Elizabeth Bruton, "From Australia to Zimmerman: A Brief History of Cable Telegraphy during World War One," unpublished manuscript, Oxford University, 2013.

54. "Trans-Atlantic Bandwidth: Then and Now."

55. Michael Sechrist, "Cyberspace in Deep Water: Protecting Undersea Communications Cables by Creating an International Public-Private Partnership," Harvard Kennedy School, 2010, 9–10.

56. See, for example, Sherry Sontag and Christopher Drew, *Blind Man's Bluff: The Untold Story of American Submarine Espionage* (New York: Public Affairs, 2016).

57. Kathleen Hicks et al., "Undersea Warfare in Northern Europe," Center for Strategic and International Studies, 2016, 11–12.

58. Dmitry Litovkin, "Top-Secret Submarine May Settle Russia's Claim in the Arctic," *Russia Beyond the Headlines*, December 13, 2012.

59. James Bamford, "Frozen Assets," *Foreign Policy*, May 11, 2015.

60. Tyler Rogoway, "Russia's Massive Arctic Research Submarine Will Be the World's Largest," *The Warzone*, May 3, 2017. http://www.thedrive.com/the-war-zone/9928/russias-massive-arctic-research-submarine-will-be-the-worlds-longest.

61. Kathleen Weinberger, "Sight Unseen—Russian Auxiliary Submarines and Asymmetric Warfare in the Undersea Domain," Center for Strategic and International Studies, March 31, 2016.

62. See, for example, "Russia's Navy Is More Rust than Ready," *War Is Boring*, August 14, 2015.

63. Laruelle, *Russia's Arctic Strategies and the Future of the Far North*, 114.

64. Tamnes, "The Significance of the North Atlantic and the Norwegian Contribution," 26.

65. Cavas, "Is Caspian Sea Fleet a Game-Changer?"

66. James Bosbotinis, "The Russian Navy in 2009: A Review of Major Developments," Corbett Paper No. 1, King's College London, April 2010, 3.

67. Jeff Lightfoot, "Mistral Mysteries," *American Interest*, December 10, 2014.

68. Ranft and Till, *The Sea in Soviet Strategy*, 154.

69. Ranft and Till, 153; and Polmar, *Guide to the Soviet Navy*, 285–86.

70. Thomas Nielsen, "Satellite Images Show Expansion of Nuclear Weapons Sites on Kola," *Barents Observer*, May 7, 2017.

71. Konyshev and Sergunin, "Russian Military Strategies in the High North," 93.

72. Trude Pettersen, "No Yachts, Only Subs," *Barents Observer*, February 22, 2016.

73. See World Health Organization, Global Health Observatory Data: Life Expectancy, 2016.

74. See U.S. Energy Information Administration, "Oil and Natural Gas Sales Accounted for 68% of Russia's Total Export Revenues in 2013," *Energy Today*, July 23, 2014.

75. See Richard Connolly and Cecilie Sendstad, "Russia's Role as an Arms Exporter: The Strategic and Economic Importance of Arms Exports for Russia," *Chatham House*, 2017.

76. Richard Connolly, "Towards Self Sufficiency? Economics as a Dimension of Russian Security and the National Security Strategy of the Russian Federation to 2020," NATO Defence College, Rome, Italy, July 2016, 2.

77. Laruelle, *Russia's Arctic Strategies and the Future of the Far North*, 118.

78. Ashley O'Keefe, "Sea Control 127—Dr. Tom Fedyszyn on Russian Navy Ops, Acquisition, and Doctrine," Center for International Maritime Security, February 1, 2017.

79. "The Russian Navy: A Historic Transition," U.S. Office of Naval Intelligence, Washington, D.C., xii–xx.

80. Alpo Juntunen, "The Baltic Sea in Russian Strategy," *Proceedings of the Royal Swedish Academy of War Sciences*, no. 4 (November 2010): 120.

81. Seth Cropsey, *Seablindness: How Political Neglect Is Choking American Seapower and What to Do About It* (New York: Encounter Books, 2017), 51.

82. Sam Lagrone, "China, Russia Kick Off Joint South China Sea Naval Exercise; Includes 'Island Seizing' Drill," U.S. Naval Institute, September 12, 2016.

83. Roald Gjeltsen, "The Role of Naval Forces in Northern Waters at the Beginning of a New Century," in *Navies in Northern Waters 1721–2000*, ed. Rolf Hobson and Tom Kristiansen (Milton Park, U.K.: Taylor and Francis, 2004), 285.

Chapter 11. The Fourth Battle of the Atlantic

1. James Foggo and Alarik Fritz, "The Fourth Battle of the Atlantic," U.S. Naval Institute *Proceedings*, June 2016.

2. Mark Ferguson, "Remarks Delivered by Adm. Mark Ferguson at the Atlantic Council," U.S. Naval Forces Europe-Africa/U.S. Sixth Fleet, October 6, 2015.

3. John Richardson, "A Design for Maintaining Maritime Superiority," U.S. Navy, 2016, 3.

4. Clive Johnstone, "NATO's Maritime Moment: A Watershed Year in Alliance Sea Power," NATO Maritime Command, January 17, 2017.

5. Michael Birnbaum, "Russian Submarines Are Prowling around Vital Undersea Cables. It's Making NATO Nervous," *Washington Post*, December 23, 2017.

6. Julianne Smith and Jerry Hendrix, "Forgotten Waters: Minding the GIUK Gap," Center for a New American Security, 2016, 7. This author also participated in the war game.

7. Kathleen Hicks, et al., "Undersea Warfare in Northern Europe," Center for Strategic and International Studies, 2016, V.

8. Julian Barnes, "A Russian Ghost Submarine, Its US Pursuers and a Deadly New Cold War," *Wall Street Journal*, October 20, 2017.

9. Norman Polmar and Edward Whitman, *Hunters and Killers*, vol. 2: *Anti-Submarine Warfare from 1943* (Annapolis, Md.: Naval Institute Press, 2016), 197.

10. See Andrew Foxall, "Close Encounters: Russian Military Activities in the Vicinity of UK Air and Sea Space, 2005–2016," Henry Jackson Society, 2017.

11. Justin Bronk, "The P-8 Poseidon for the UK," Commentary, Royal United Services Institute, July 14, 2016.

12. Shawn Snow, "The Corps' Largest Rotational Presence to Norway Kicks Off," *Marine Corps Times*, October 1, 2018.

13. Ine Eriksen Soreide, "Minister of Defense Speech at ACUS Conference 'Charting NATO's Future,'" Government of Norway, September 25, 2015.

14. Havard Klevberg, *Request Tango: 333 skvadron pa ubatsjakt—maritime luftoperasjoner i norsk sikkerhetspolitikk* (Oslo: Universitetsforlaget, 2012), 379–90.

15. Christopher Cavas, "Resurgent Russia Drawing Northern Nations Closer," *Defense News*, September 8, 2015.

16. Shawn Snow, "US Plans $200 Million Build-Up of European Air Bases Flanking Russia," *Air Force Times*, December 17, 2017.

17. Icelandic Coast Guard, Reykjavik, Iceland, briefing to the author, November 15, 2016.

18. Trude Pettersen, "Norway Improving Infrastructure on Jan Mayen," *Barents Observer*, August 12, 2015.

19. Chris Cope, "Unst Radar Base Work to Begin in October," *Shetland News*, September 16, 2017.

20. Mark Faram, "Back to the Future with 2nd Fleet," *Navy Times*, August 24, 2018.

21. Franklin D. Kramer and Hans Binnendijk, "Meeting the Russian Conventional Challenge: Effective Deterrence by Prompt Reinforcement," Atlantic Council, February 2018.

22. Carol Redfield, "The Seaport of Debarkation: A Critical Vulnerability for the Operational Commander?," U.S. Naval War College, February 3, 2003, 3.

23. Redfield, 7–8.

24. U.S. Army Europe, "Freedom of Movement Assessment: Moving Towards a Military Schengen Zone," USEUCOM, March 15, 2017.

25. Michael Johnson and Brent Coryell, "Logistics Forecasting and Estimates in the Brigade Combat Team," *Army Sustainment*, November 2016.

26. See William Pagonis, *Moving Mountains: Lessons in Leadership and Logistics from the Gulf War* (Cambridge, Mass.: Harvard University Press, 1994).

27. Jen Judson, "US Army May Send Larger Deployments to Europe," *Inside Defense*, December 14, 2017.

28. United States Army Public Affairs Europe, "10th CAB Equipment Begins Arriving in Europe," U.S. Army, February 8, 2017.

29. Judson, "US Army May Send Larger Deployments to Europe."

30. Scott Wyland, "Navy's Civilian-Run Ships Playing Bigger Supply Role Amid Russia Tensions," *Stars and Stripes*, February 26, 2018.

Chapter 12. The Contested North Atlantic in the Twenty-First Century

1. "Navy Announces Launch of Task Force Ocean, Plans to Advance Ocean Science," press release, Office of the Oceanographer of the U.S. Navy, March 24, 2017.

2. "Charting the Arctic Sea's Changing Environment," press release, Allied Command Transformation, July 2017.

3. See U.S. Department of Defense, "The National Defense Strategy of the United States of America," Washington, D.C., 2017.

4. David Larter, "The New Navy Secretary Is Inheriting a Mess: Here's How the Navy Wants to Fix It," *Navy Times*, February 4, 2017; and Bryan Clark and Jesse Sloman, "Deploying Beyond Their Means: America's Navy and Marine Corps at a Tipping Point," Center for Strategic and Budgetary Assessments, November 2015.

5. David Larter, "Navy Crews at Fault in Fatal Collisions, Investigation Finds," *Defense News*, November 1, 2017.

6. John Schaus, Lauren Dickey, and Andrew Metrick, "Asia's Looming Sub-Surface Challenge," *War on the Rocks*, August 11, 2016.

7. Jane Perlez, "Panetta Outlines New Weaponry for Pacific," *New York Times*, June 1, 2012.

8. Colin McInnes, "British Sea Power at the Millennium," *Comparative Strategy* 17, no. 2: 134.

9. See Elizabeth Wishnick, *China's Interests and Goals in the Arctic: Implications for the United States* (Carlisle, Pa.: US Army War College Strategic Studies Institute,

March 2017); and Shilo Rainwater, "Race to the North: China's Arctic Strategy and Its Implications," *Naval War College Review* (Spring 2013): 62–82.

10. Magnus Nordenman, "China and Russia's Joint Sea 2017 Baltic Naval Exercise Highlight a New Normal in Europe," United States Naval Institute, July 5, 2017.

11. Andrew Higgins, "A Rare Arctic Land Sale Stokes Worry in Norway," *New York Times*, September 27, 2014.

12. Lasse Eriksen, "Norway Says No to Chinese Radar on Svalbard," *Scandasia*, September 15, 2014.

13. Teis Jensen, "Greenland Shortlists Chinese Company for Airport Construction Despite Denmark's Concerns," *Reuters*, March 27, 2018.

14. Aaron Mehta, "How a Potential Chinese-Built Airport in Greenland Could Be Risky for a Vital U.S. Air Base," *Defense News*, September 7, 2018.

15. Michael Rubin, "Don't Give China an Atlantic Base," *Commentary*, November 29, 2015.

16. Ting Shi, "Portugal Open to China Investment in Azores as US Sway Wanes," Bloomberg, October 11, 2016.

17. Charles Emmerson, *The Future History of the Arctic* (New York: Public Affairs, 2010), 263–74.

18. See Alyson Bailes, "Scotland as an Independent Small State: Where Would It Seek Shelter?," *Stjornmal & Stjornsysla*, no. 1 (2013): 1–20.

19. Dan De Luce, "Britain Needs a New Place to Park Its Nukes," *Foreign Policy*, July 10, 2016.

20. Daniel Gros, "Revisiting Sanctions on Russia and Counter-Sanctions on the EU: The Economic Impact Three Years Later," Center for European Policy Studies, July 13, 2014.

21. Anton Troianovski, "Faroe Islands Boom by Selling Salmon to Russia," *Wall Street Journal*, February 20, 2015.

22. See Beinta Jakupsstovu, "The Faroe Island's Security Policy in a Process of Devolution," *Stjornmal & Stjornsysla*, no. 2 (2012): 413–30.

23. "Resistance Grows in EU to New Russia Sanctions," *Deutsche Welle*, September 5, 2014.

24. Rachel Nuwer, "A Soviet Ghost Town in the Arctic, Pyramiden Stands Alone," *Smithsonian*, May 19, 2014.

25. "Norway in Arctic Dispute with Russia Over Rogozin Visit," *BBC News*, April 20, 2015.

26. James Stavridis, *The Accidental Admiral: A Sailor Takes Command at NATO* (Annapolis, Md.: Naval Institute Press, 2014), 82, 101.

27. Jim Wolf, "China Key Suspect in US Satellite Hacks: Commission," *Reuters*, October 28, 2011.

28. Richard Connolly, "Towards Self Sufficiency? Economics as a Dimension of Russian Security and the National Security Strategy of the Russian Federation to 2020," NATO Defence College, Rome, Italy, July 2016, 5.

29. Martin Murphy, Frank Hoffman, and Gary Schaub, "Hybrid Maritime Warfare and the Baltic Sea Region," Centre for Military Studies, 2016, 5.

30. Emmerson, *The Future History of the Arctic*, 103.

31. See, for example, Peter Dombrowski and Chris Demchak, "Cyber War, Cybered Conflict, and the Maritime Domain," *Naval War College Review* (March 2014): 71–96.

32. Primetrica, Inc, "Submarine Cable Map," https://www.submarinecablemap.com/.

33. Michael Byrne, "Hackers Launch All-Out Assault on Norway's Oil and Gas Industry," *Motherboard*, August 31, 2014.

34. Atle Staalesen, "FSB Trains Counter-Terrorism at Arctic Oil Installation," *Barents Observer*, June 17, 2014.

35. "Russia's Fake 'Electronic Bomb,'" Atlantic Council, May 8, 2016.

36. Matthew Krull, "Foreign Disinformation Is a Threat to Military Readiness, Too," *Defense One*, February 16, 2018.

37. Milan Vego, "Patrolling the Deep: Critical Anti-Submarine Warfare Skills Must Be Restored," *Armed Forces Journal* (September 2008): 24–39.

38. William Broad, "Scientists Fight Navy Plan to Shut Far-Flung Undersea Spy System," *New York Times*, July 12, 1994.

39. "National Security Implications of Climate Changes for US Naval Forces," National Research Council, 2011, 107–12.

40. "The Future of Anti-Submarine Warfare," NATO Public Affairs video presentation, June 3, 2015.

41. "Nearly Undetectable Submarines Threaten US Ships and Shipping," *Leidos*, July 20, 2016.

42. Bradley Peniston, "The Future of the US Navy," *Defense One*, May 2017, 10.

43. "Repositioning Anti-Submarine Warfare," Allied Command Transformation press release, July 2017.

44. Beth Stevenson, "DSEI: Northrop Maintains Hope in UK, Norway Markets for Triton," *FlightGlobal*, September 15, 2015.

45. Norman Polmar and Edward Whitman, *Hunters and Killers,* vol. 2: *Anti-Submarine Warfare from 1943* (Annapolis, Md.: Naval Institute Press, 2016), 198.

46. George Galdorisi, "We Need an Undersea Constellation," U.S. Naval Institute *Proceedings* 141 (June 2015).

47. Shelby Sullivan, "Mobile Offboard Clandestine Communications and Approach (MOCCA)," Defense Advanced Research Projects Agency, n.d.

48. Steve Lohr, "The Age of Big Data," *New York Times*, February 11, 2012.

49. Sydney Freedberg, "Transparent Sea: The Unstealthy Future of Submarines," *Breaking Defense*, January 22, 2015.

50. Bryan Clark, "The Emerging Era in Undersea Warfare," Center for Strategic and Budgetary Assessments, 2015, 16.

51. Vego, "Patrolling the Deep," 24–39; and Norman Friedman, "Finding Submerged Submarines—with Lasers," U.S. Naval Institute *Proceedings* 119 (March 1993).

52. Julianne Smith and Jerry Hendrix, "Forgotten Waters: Minding the GIUK Gap," Center for New American Security, 5.

53. Keith Johnson, "US Falls Behind in Arctic Great Game," *Foreign Policy*, May 24, 2016.

54. Alec Luhn, "Arctic Cities Crumble As Climate Change Thaws Permafrost," *Wired*, October 20, 2016.

55. Andreas Kuersten, "Icebreakers and US Power: Separating Fact from Fiction," *War on the Rocks*, October 11, 2016.

56. Magnus Nordenman, "The Russian Challenge in the Arctic Isn't About Icebreakers," *Defense News*, February 24, 2017.

Chapter 13. NATO and the United States in the Twenty-First-Century North Atlantic

1. Magnus Nordenman, "Back to the North: The Future of the German Navy in the New European Security Environment," Atlantic Council, 2017.

2. Magnus Nordenman, "NATO's Next Consortium: Maritime Patrol Aircraft," Atlantic Council, June 2016.

3. Peter Hudson and Peter Roberts, "The UK and the North Atlantic: A Military Perspective," in *NATO and the North Atlantic: Revitalising Collective Defense*, ed. John Andreas Olsen (London: Royal United Services Institute, 2017), 90.

4. "NATO Allies Move to Replace Aging Maritime Anti-Submarine and Patrol Aircraft Capabilities," press release, NATO, July 7, 2017.

5. See Magnus Nordenman and Franklin D. Kramer, "A Maritime Framework for the Baltic Sea," Atlantic Council, 2016.

6. See Mark Gunziger et al., "Force Planning for the Era of Great Power Competition," Center for Strategic and Budgetary Assessments, 2017; and Bryan Clark et al., "Restoring American Seapower: A New Fleet Architecture for the United States Navy," Center for Strategic and Budgetary Assessments, 2017, 46–48.

7. Gunziger, et al., "Force Planning for the Era of Great Power Competition," 56–59.

8. Steve Wills, "A New Gap in the High North and Forward Defense Against Russian Naval Power," Center for International Maritime Security, July 17, 2018.

9. Gunziger, et al., "Force Planning for the Era of Great Power Competition," 55.

10. Chris Parry, *Super Highway: Sea Power in the 21st Century* (London: Elliott and Thompson, 2014), 170–71.

11. See James Stavridis, "Maritime Hybrid Warfare Is Coming," *Proceedings* 142 (December 2016).

Conclusion

1. See, for example, Richard Shirreff, *War with Russia: An Urgent Warning from Senior Military Command* (Quercus, 2016).

Selected Bibliography

Books

Archer, Clive, ed. *The Soviet Union and Northern Waters*. London: Royal Institute of International Affairs, 1988.

Atkison, Rick. *An Army at Dawn: The War in North Africa, 1942–1943*. New York: Holt, 2007.

Boot, Max. *The Savage Wars of Peace: Small Wars and the Rise of American Power*. New York: Basic Books, 2014.

Bruton, Elizabeth. "From Australia to Zimmerman: A Brief History of Cable Telegraphy during World War One." Unpublished manuscript, Oxford University, 2013.

Burleson, Clyde. *Kursk Down: The Shocking True Story of the Sinking of a Russian Nuclear Submarine*. New York: Warner Books, 2002.

Clancy, Tom. *Carrier: A Guided Tour of an Aircraft Carrier*. New York: Berkley Books, 1999.

Cropsey, Seth. *Seablindness: How Political Neglect Is Choking American Seapower and What to Do About It*. New York: Encounter Books, 2017.

Draper, Alan. *European Defense Equipment Collaboration: Britain's Involvement, 1957–87*. Berlin: Springer, 1990.

Emmerson, Charles. *The Future History of the Arctic*. New York: Public Affairs, 2010.

Gjorv, Gunhild Hoogensen, ed. *Environmental and Human Security in the Arctic*. London: Routledge, 2013.

Hastings, Max, and Simon Jenkins. *The Battle for the Falklands*. New York: W. W. Norton & Company, 1984.

Heininen, Lassi, ed. *Security and Sovereignty in the North Atlantic*. Basingstoke, U.K.: Palgrave Macmillan, 2014.

Herman, Arthur. *Freedom's Forge: How American Business Produced Victory in World War II*. New York: Random House, 2013.

Hill, Fiona, and Clifford Gaddy. *Mr. Putin: Operative in the Kremlin*. Washington, D.C.: Brookings Institution, 2013.

Hobson, Rolf, and Tom Kristiansen, eds. *Navies in Northern Waters 1721–2000*. Milton Park, U.K.: Taylor and Francis, 2004.

Jordan, Robert. *Alliance Strategy and Navies: The Evolution and Scope of NATO's Maritime Dimension*. New York: St. Martin's Press, 1990.

Keegan, John. *The Price of Admiralty*. London: Hutchison, 1988.

Klevberg, Havard. *Request Tango: 333 skvadron pa ubatsjakt—maritime luftoperasjoner i norsk sikkerhetspolitikk*. Oslo: Universitetsforlaget, 2012.

Krupnick, Charles. *Decommissioned Russian Nuclear Submarines and International Cooperation*. Jefferson, N.C.: McFarland & Company, 2001.

Larson, Erik. *Dead Wake: The Last Crossing of the Lusitania*. New York: Crown Publishing, 2015.

Laruelle, Marlene. *Russia's Arctic Strategies and the Future of the Far North*. Abingdon, U.K.: Routledge, 2014.

Lehman, John. *Oceans Ventured: Winning the Cold War at Sea*. New York: W. W. Norton & Company, 2018.

Maynard, Charles. *The Murmansk Venture*. New York: Arno Press, 1971.

More, Robert. *A Time to Die*. New York: Random House, 2003.

Olsen, John Andreas, ed. *NATO and the North Atlantic: Revitalising Collective Defense*. London: Royal United Services Institute, 2017.

Pagonis, William. *Moving Mountains: Lessons in Leadership and Logistics from the Gulf War*. Cambridge, Mass.: Harvard University Press, 1994.

Parry, Chris. *Super Highway: Sea Power in the 21st Century*. London: Elliott and Thompson, 2014.

Polmar, Norman. *Guide to the Soviet Navy*. Annapolis, Md.: Naval Institute Press, 1986.

Polmar, Norman, and Edward Whitman. *Hunter and Killers: Volume 2: Anti-Submarine Warfare Since 1943*. Annapolis, Md.: Naval Institute Press, 2016.

Ranft, Bryan, and Geoffrey Till. *The Sea in Soviet Strategy*. Annapolis, Md.: Naval Institute Press, 1983.

Rieff, David. *Slaughterhouse: Bosnia and the Failure of the West*. New York: Touchstone, 1996.

Rodriguez, Luis, and Sergiy Glebov. *Military Bases: Historical Perspectives, Contemporary Challenges*. Amsterdam: IOS Press, 2009.

Scahill, Jeremy. *The Assassination Complex: Inside the Government's Secret Drone Warfare Program*. New York: Simon & Schuster, 2016.

Schank, John F., et al. *Learning from Experience*. Vol. 4: *Lessons from Australia's Collins Submarine Program*. Arlington, Va.: RAND Corporation, 2011.

Singer, P. W. *Wired for War: The Robotics Revolution and Conflict in the 21st Century*. New York: Penguin Press, 2009.

Skogan, John, and Arne Brundtland. *Soviet Sea Power in Northern Waters*. New York: St. Martin's Press, 1990.

Sontag, Sherry, and Christopher Drew. *Blind Man's Bluff: The Untold Story of American Submarine Espionage*. New York: Public Affairs, 2016.

Stavridis, James. *The Accidental Admiral: A Sailor Takes Command at NATO.* Annapolis, Md.: Naval Institute Press, 2014.

——. *Sea Power: The History and Geopolitics of the World's Oceans.* New York: Penguin Books, 2017.

Tangredi, Sam. *Anti-Access Warfare: Countering A2/AD Strategies.* Annapolis, Md.: Naval Institute Press, 2013.

Truscott, Peter. *Kursk.* New York: Simon & Schuster, 2002.

Weir, Gary, and Sandra J. Doyle, eds. *You Cannot Surge Trust: Combined Naval Operations of the Royal Australian Navy, Canadian Navy, Royal Navy, and United States Navy, 1991–2003.* Washington, D.C.: Naval History and Heritage Command, 2015.

Wells, Anthony. *A Tale of Two Navies: Geopolitics, Technology, and Strategy in the United States Navy and the Royal Navy, 1960–2015.* Annapolis, Md.: Naval Institute Press, 2017.

Winchester, Simon. *Atlantic: Great Sea Battles, Heroic Discoveries, Titanic Storms, and a Vast Ocean of a Million Stories.* New York: HarperCollins, 2011.

Winkler, David. *Cold War at Sea: High-Seas Confrontations between the United States and the Soviet Union.* Annapolis, Md.: Naval Institute Press, 2000.

Winkler, Jonathan Reed. *Nexus: Strategic Communications and American Security in World War I.* Cambridge, Mass.: Harvard University Press, 2008.

Wishnick, Elizabeth. *China's Interests and Goals in the Arctic: Implications for the United States.* Carlisle, Pa.: U.S. Army War College Strategic Studies Institute, March 2017.

Journal Articles

Allard, Dean. "Strategic Views of the US Navy and NATO on the Northern Flank 1917–1991." *Northern Mariner* 11, no. 1 (January 2001).

Altmann, John. "Russian A2/AD in the Eastern Mediterranean: A Growing Risk." *Naval War College Review* 69, no. 1 (Winter 2016).

Bremmer, Jan. "Defeating the U-Boat: Inventing Anti-Submarine Warfare." *Naval War College Newport Papers* 36 (2010).

Cote, Owen. "The Third Battle: Innovation in the US Navy's Silent Cold War Struggle with Soviet Submarines." *Naval War College Newport Papers* 16 (2003).

Davenport, Tara. "Submarine Cables, Cybersecurity and International Law: An Intersectional Analysis." *Catholic University Journal of Law and Technology* 24, no. 1 (December 2015).

Dombrowski, Peter, and Chris C. Demchak. "Cyber War, Cybered Conflict, and the Maritime Domain." *Naval War College Review* (March 2014).

Dorfer, Ingemar. "Kola Has Lost Significance." U.S. Naval Institute *Proceedings* 128 (March 2002).

Foggo, James, and Alarik Fritz. "The Fourth Battle of the Atlantic." U.S. Naval Institute *Proceedings* (June 2016).

Hattendorf, John. "The Evolution of the US Navy's Maritime Strategy, 1977–1986." *Naval War College Newport Papers* 19 (2004).

Hooker, Richard. "NATO's Northern Flank: A Critique of the Maritime Strategy." *Parameters* (June 1989).

Ingimundarson, Valur. "Fighting the Cod Wars in the Cold War: Iceland's Challenge to the Western Alliance in the 1970s." *RUSI Journal* 148, no. 3. (2008).

Jakupsstovu, Beinta. "The Faroe Islands' Security Policy in a Process of Devolution." *Stjornmal & Stjornsysla*, no. 2 (2012).

Kucik, Cory. "NATO's Future: It's Far from the North Atlantic." U.S. Naval Institute *Proceedings* 140 (September 2014).

Lightfoot, Jeff. "Mistral Mysteries." *American Interest*, December 10, 2014.

Nordenman, Magnus. "The Incredible, Shrinking Modern Military." *Atlantic Monthly*, November 12, 2012.

Osthagen, Andreas. "High North, Low Politics—Maritime Cooperation with Russia in the Arctic." *Arctic Review on Law and Politics* 7, no. 1 (2016).

Rainwater, Shilo. "Race to the North: China's Arctic Strategy and Its Implications." *Naval War College Review* (Spring 2013).

Stavridis, James. "Maritime Hybrid Warfare Is Coming" U.S. Naval Institute *Proceedings* 142 (December 2016).

Weigert, Hans W. "Iceland, Greenland and the United States." *Foreign Affairs* 23, no. 1 (October 1944).

Whitby, Michael. "Odd Jobs: Canada's Use of Submarines on Fisheries Patrols 1993–1995. Part I." *Submarine Review* (September 2017).

Willett, Lee. "Game Changer: Russian Sub-Launched Cruise Missiles Bring Strategic Effect." *Jane's International Defence Review*, April 27, 2017.

Reports

Barrett, Mark, et al. "Assured Access to the Global Commons." Allied Command Transformation, 2010.

Binnendijk, Hans, and Franklin D Kramer. "Meeting the Russian Conventional Challenge: Effective Deterrence by Prompt Reinforcement." Atlantic Council, February 2018.

Bosbotinis, James. "The Russian Navy in 2009: A Review of Major Developments." Corbett Paper No. 1, King's College London, April 2010.

Brands, Hal. "Dealing with Allies in Decline." Center for Strategic and Budgetary Assessments, 2017.

Clark, Bryan. "The Emerging Era in Undersea Warfare." Center for Strategic and Budgetary Assessments, 2015.

Clark, Bryan, et al. "Restoring American Seapower: A New Fleet Architecture for the United States Navy." Center for Strategic and Budgetary Assessments, 2017.

Connolly, Richard. "Towards Self Sufficiency? Economics as a Dimension of Russian Security and the National Security Strategy of the Russian Federation to 2020." NATO Defence College, Rome, Italy, July 2016.

Connolly, Richard, and Cecilie Sendstad. "Russia's Role as an Arms Exporter: The Strategic and Economic Importance of Arms Exports for Russia." *Chatham House*, 2017.

Foxall, Andrew. "Close Encounters: Russian Military Activities in the Vicinity of UK Air and Sea Space, 2005–2016." Henry Jackson Society, 2017.

Frear, Thomas. "Anatomy of a Russian Exercise." European Leadership Network, August 12, 2015.

Gunzinger, Mark, et al. "Force Planning for the Era of Great Power Competition." Center for Strategic and Budgetary Assessments, 2017.

Harkavy, Robert. "Bases Abroad: The Global Foreign Military Presence." Stockholm International Peace Research Institute, 1989.

Hendrix, Jerry. "Retreat from Range: The Rise and Fall of Carrier Aviation." Center for a New American Security, 2015.

Hicks, Kathleen, et al. "Undersea Warfare in Northern Europe." Center for Strategic and International Studies, 2016.

Kopte, Susanne. "Nuclear Submarine Decommissioning and Related Problems." Bonn International Center for Conversion, 1997.

Larrabee, F. Stephen, et al. "NATO and the Challenges of Austerity." RAND Corporation, 2012.

McDermott, Roger, and Tor Bukkvoll. "Russia in the Precision-Strike Regime: Military Theory, Procurement and Operational Impact." Norwegian Defense Research Establishment, 2017.

McGrath, Bryan. "NATO at Sea: Trends in Allied Seapower." American Enterprise Institute, 2013.

Metrick, Andrew, and Kathleen Hicks. "Contested Seas: Maritime Domain Awareness in Northern Europe." Center for Strategic and International Studies, 2018.

Murphy, Martin, Gary Schaub, and Frank Hoffman. "Hybrid Maritime Warfare and the Baltic Sea Region." Center for Military Studies, University of Copenhagen, November 2016.

Nicholls, David. "Cruise Missiles and Modern War: Strategic and Technological Implications." Center for Strategy and Technology, Air War College, Occasional Paper No. 13.

Nordenman, Magnus. "Back to the North: The Future of the German Navy in the New European Security Environment." Atlantic Council, 2017.

———. "NATO's Next Consortium: Maritime Patrol Aircraft." Atlantic Council, June 2016.

Nordenman, Magnus, and Franklin D. Kramer. "A Maritime Framework for the Baltic Sea." Atlantic Council, 2016.

Perkins, William. "Alliance Airborne Anti-Submarine Warfare." Joint Air Power Competency Center, June 2016.

Poe, Stacey. "Rules of Engagement: Complexities of Coalition Interactions in Operations Other Than War." Defense Technical Information Center, 1995.

Redfield, Carol. "The Seaport of Debarkation: A Critical Vulnerability for the Operational Commander?" U.S. Naval War College, February 3, 2003.

"Russian Military Power: Building a Military to Support Great Power Aspirations." United States Defense Intelligence Agency, 2017.

"The Russian Navy: A Historic Transition." U.S. Office of Naval Intelligence, Washington, D.C.

Smith, Julianne, and Jerry Hendrix. "Forgotten Waters: Minding the GIUK Gap." Center for New American Security, 2017.

Smith-Windsor, Brooke, "NATO's Maritime Strategy and the Libya Crisis as Seen from the Sea." NATO Defense College, Rome, 2013.

Stillion, John, and Bryan Clark. "What It Takes to Win: Succeeding in 21st Century Battle Network Competitions." Center for Strategic and Budgetary Assessments, 2015.

Swartz, Peter. "Understanding an Adversary's Strategic and Operational Calculus: A Late Cold War Case Study with 21st Century Applicability." Center for Naval Analyses, 2013.

U.S. Department of Defense. "The National Defense Strategy of the United States of America." Washington, D.C., 2017.

U.S. Office of Technology Assessment. "Nuclear Wastes in the Arctic: An Analysis of Arctic and Other Regional Impacts from Soviet Nuclear Contamination." Washington, D.C.: Government Printing Office, 1995.

Watts, Barry D. "The Maturing Revolution in Military Affairs." Center for Strategic and Budgetary Assessments, 2011.

Wills, Steve. "A New Gap in the High North and Forward Defense against Russian Naval Power." Center for International Maritime Security, July 17, 2018.

INDEX

About the Author

Magnus Nordenman is a noted NATO and naval affairs expert and has served as the director of the Transatlantic Security Initiative and the deputy director of the Brent Scowcroft Center on International Security at the Atlantic Council in Washington, D.C. He speaks and writes on European and transatlantic security, NATO in the maritime domain, defense and security in Northern Europe, NATO transformation, and U.S.-European relations. He has provided advice and insights to a range of departments and agencies, including the U.S. Department of Defense, Department of State, and armed forces, as well as to NATO and the ministries of defense and foreign affairs of U.S. friends, allies, and partners in Europe. His commentary and insights have been featured by, among others, the *New York Times*, the BBC, MSNBC, Al Jazeera, *Defense News*, *Defense One*, and the U.S. Naval Institute. Before his time at the Atlantic Council, Nordenman served as a defense analyst with a small Washington-based consulting company and as a consultant to the defense industry. A proud American immigrant, he arrived in the United States in 1998. He is a graduate of the Virginia Military Institute and earned his MA in national security studies from the Patterson School at the University of Kentucky. He lives with his family in Northern Virginia.